MEMOIRS
of the
American Mathematical Society

Number 836

The Second Duals of Beurling Algebras

H. G. Dales
A. T.-M. Lau

September 2005 • Volume 177 • Number 836 (end of volume) • ISSN 0065-9266

American Mathematical Society
Providence, Rhode Island

2000 *Mathematics Subject Classification.*
Primary 43A10, 43A20; Secondary 46J10.

Library of Congress Cataloging-in-Publication Data

Dales, H. G. (Harold G.), 1944–
The second duals of Beurling algebras / H. G. Dales, A. T.-M. Lau.
p. cm. — (Memoirs of the American Mathematical Society, ISSN 0065-9266 ; no. 836)
"Volume 177, number 836 (end of volume)."
Includes bibliographical references and index.
ISBN 0-8218-3774-5 (alk. paper)
1. Banach algebras. 2. Measure algebras. 3. Topological groups. I. Lau, Anthony To-Ming. II. Title. III. Series.

QA3.A57 no. 836
[QA326]
510 s—dc22
[512′.554] 2005048020

Memoirs of the American Mathematical Society

This journal is devoted entirely to research in pure and applied mathematics.

Subscription information. The 2005 subscription begins with volume 173 and consists of six mailings, each containing one or more numbers. Subscription prices for 2005 are $606 list, $485 institutional member. A late charge of 10% of the subscription price will be imposed on orders received from nonmembers after January 1 of the subscription year. Subscribers outside the United States and India must pay a postage surcharge of $31; subscribers in India must pay a postage surcharge of $43. Expedited delivery to destinations in North America $35; elsewhere $130. Each number may be ordered separately; *please specify number* when ordering an individual number. For prices and titles of recently released numbers, see the New Publications sections of the *Notices of the American Mathematical Society.*

Back number information. For back issues see the *AMS Catalog of Publications.*

Subscriptions and orders should be addressed to the American Mathematical Society, P. O. Box 845904, Boston, MA 02284-5904, USA. *All orders must be accompanied by payment.* Other correspondence should be addressed to 201 Charles Street, Providence, RI 02904-2294, USA.

Memoirs of the American Mathematical Society is published bimonthly (each volume consisting usually of more than one number) by the American Mathematical Society at 201 Charles Street, Providence, RI 02904-2294, USA. Periodicals postage paid at Providence, RI. Postmaster: Send address changes to Memoirs, American Mathematical Society, 201 Charles Street, Providence, RI 02904-2294, USA.

This publication is indexed in *Science Citation Index*®, *SciSearch*®, *Research Alert*®, *CompuMath Citation Index*®, *Current Contents*®/*Physical, Chemical & Earth Sciences.*
Printed in the United States of America.

The paper used in this book is acid-free and falls within the guidelines established to ensure permanence and durability.
Visit the AMS home page at http://www.ams.org/

10 9 8 7 6 5 4 3 2 1 10 09 08 07 06 05

Contents

Chapter 1. Introduction 1

Chapter 2. Definitions and Preliminary Results 7

Chapter 3. Repeated Limit Conditions 25

Chapter 4. Examples 35

Chapter 5. Introverted Subspaces 45

Chapter 6. Banach Algebras of Operators 53

Chapter 7. Beurling Algebras 65

Chapter 8. The Second Dual of $\ell^1(G,\omega)$ 95

Chapter 9. Algebras on Discrete, Abelian Groups 111

Chapter 10. Beurling Algebras on $\mathbb{F}_2$ 131

Chapter 11. Topological Centres of Duals of Introverted Subspaces 141

Chapter 12. The Second Dual of $L^1(G,\omega)$ 153

Chapter 13. Derivations into Second Duals 167

Chapter 14. Open Questions 175

Bibliography 177

Index 185

Index of Symbols 189

Abstract

Let A be a Banach algebra, with second dual space A''. We propose to study the space A'' as a Banach algebra. There are two Banach algebra products on A'', denoted by $\Box$ and $\Diamond$. The Banach algebra A is *Arens regular* if the two products $\Box$ and $\Diamond$ coincide on A''. In fact, A'' has two *topological centres* denoted by $\mathfrak{Z}_t^{(1)}(A'')$ and $\mathfrak{Z}_t^{(2)}(A'')$ with $A \subset \mathfrak{Z}_t^{(j)}(A'') \subset A''$ $(j = 1, 2)$, and A is Arens regular if and only if $\mathfrak{Z}_t^{(1)}(A'') = \mathfrak{Z}_t^{(2)}(A'') = A''$. At the other extreme, A is *strongly Arens irregular* if $\mathfrak{Z}_t^{(1)}(A'') = \mathfrak{Z}_t^{(2)}(A'') = A$. We shall give many examples to show that these two topological centres can be different, and can lie strictly between A and A''.

We shall discuss the algebraic structure of the Banach algebra $(A'', \Box)$; in particular, we shall seek to determine its radical and when this algebra has a strong Wedderburn decomposition. We are also particularly concerned to discuss the algebraic relationship between the two algebras $(A'', \Box)$ and $(A'', \Diamond)$.

Most of our theory and examples will be based on a study of the weighted Beurling algebras $L^1(G, \omega)$, where ω is a weight function on the locally compact group G. The case where G is discrete and the algebra is $\ell^1(G, \omega)$ is particularly important. We shall also discuss a large variety of other examples. These include a weight ω on $\mathbb{Z}$ such that $\ell^1(\mathbb{Z}, \omega)$ is neither Arens regular nor strongly Arens irregular, and such that the radical of $(\ell^1(\mathbb{Z}, \omega)'', \Box)$ is a nilpotent ideal of index exactly 3, and a weight ω on $\mathbb{F}_2$ such that two topological centres of the second dual of $\ell^1(\mathbb{F}_2, \omega)$ may be different, and that the radicals of the two second duals may have different indices of nilpotence.

2000 *Mathematics Subject Classification.* Primary 43A10, 43A20; secondary 46J10.

Key words and phrases. Banach algebras, topological centres, Arens products, Arens regular, strongly Arens irregular, radical, strong Wedderburn decomposition, nilpotent ideals, locally compact group, group algebra, weight function, Beurling algebra, repeated limit, measure algebra, introverted spaces, free group, derivations, amenable, 2-weakly amenable.

We acknowledge with thanks the financial support of NSERC grant A7679.

Received by the editor June 6, 2003.

CHAPTER 1

Introduction

The purpose of this memoir is to study the second duals of a Banach algebra A. There are two such duals, each itself a Banach algebra, for which we use the notations $(A'', \Box)$ and $(A'', \Diamond)$. The Banach algebra A is said to be *Arens regular* if the two products $\Box$ and $\Diamond$ coincide on A''. In fact, A'' has two *topological centres* denoted by $\mathfrak{Z}_t^{(1)}(A'')$ and $\mathfrak{Z}_t^{(2)}(A'')$ with

$$A \subset \mathfrak{Z}_t^{(j)}(A'') \subset A'' \quad (j = 1, 2)$$

(where we regard A as a subspace of A''), and A is Arens regular if and only if $\mathfrak{Z}_t^{(1)}(A'') = \mathfrak{Z}_t^{(2)}(A'') = A''$. At the other extreme, A is *strongly Arens irregular* if $\mathfrak{Z}_t^{(1)}(A'') = \mathfrak{Z}_t^{(2)}(A'') = A$. In the case where A is commutative, $\mathfrak{Z}_t^{(1)}(A'') = \mathfrak{Z}_t^{(2)}(A'') = \mathfrak{Z}(A'')$, the centre of both of the algebras $(A'', \Box)$ and $(A'', \Diamond)$. We shall give examples of many non-commutative Banach algebras to show that these two topological centres can be different, and that they can be equal to A, to A'', and to certain strictly intermediate closed subalgebras of A''.

We shall discuss the algebraic structure of the Banach algebra $(A'', \Box)$; in particular, we shall seek to determine its radical and when this algebra has a strong Wedderburn decomposition.

We are particularly concerned to discuss the relationship between the two algebras $(A'', \Box)$ and $(A'', \Diamond)$: for example, we shall note in Example 6.2 that the (Jacobson) radicals of the two algebras are not necessarily the same set, and in Theorem 10.12 that the two radicals can be the same set, but have different orders of nilpotence.

Finally, we shall also study the class of continuous derivations from A to A'', when A'' is regarded as a Banach A-bimodule.

We shall throughout exemplify our general theory by studying the two Banach algebras $(A'', \Box)$ and $(A'', \Diamond)$ in the case where A is a weighted convolution algebra $L^1(G, \omega)$: here G is a locally compact group, not necessarily abelian, and ω is a weight function on G. The case where G is discrete and we are considering the Banach algebra $\ell^1(G, \omega)$ is particularly important. However we shall also consider a considerable number of other examples.

The pioneering work on what are now called the Arens products on the second dual A'' of a Banach algebra A is that of Richard Arens, more than half a century ago (see [Ar1], [Ar2]). Later, in a seminal paper of 1961, Civin and Yood [CiY] concentrated on the special case where A is the group algebra $L^1(G)$ of a locally compact group G, proving that, in the case where G is abelian, $L^1(G)$ is Arens regular only if G is finite; this was established for general groups G by Young in [Y2]. It was finally proved in 1988 that $L^1(G)$ is strongly Arens irregular for each locally compact group G [LLos1]. (This was proved earlier for compact groups G in [IPyU]; see also [BaLPy].) We are indebted to Craw and Young [CrY] for the determination when weighted group algebras are Arens regular.

The theory of Arens products on the second duals of Banach algebras is described in the texts [Pa2] and [D]. Our memoir builds on these foundations.

Although most of our results are new, we have sought to survey the known theory, and in a few cases we have repeated proofs that already exist in the literature.

In Chapter 2, we shall recall the background in Banach algebra theory that we shall require, and introduce the key concepts of the two Arens products $\Box$ and $\Diamond$ on the second dual A'' of a Banach algebra A; we shall also define the two topological centres $\mathfrak{Z}_t^{(1)}(A'')$ and $\mathfrak{Z}_t^{(2)}(A'')$ in A''. This leads to our definitions of Arens regularity and of (left and right) strong Arens irregularity.

Grothendieck's 'repeated limit condition' plays an important role in the study of Arens regularity. In Chapter 3, we shall describe this condition and establish some results in the form that we shall use them; we shall also recall the definition of almost periodic and weakly almost periodic elements in the dual space A' of a Banach algebra A.

In Chapter 4, we shall give a variety of examples of Banach algebras A and of their two second dual algebras. In particular, we shall recall the properties of C^*-algebras A, for which $\mathfrak{Z}_t^{(1)}(A'') = \mathfrak{Z}_t^{(2)}(A'') = A''$, and of group algebras A, for which $\mathfrak{Z}_t^{(1)}(A'') = \mathfrak{Z}_t^{(2)}(A'') = A$. We shall also collect a substantial number of specific examples, most already known.

In Chapter 5, we shall introduce the class of (left-) introverted subspaces X of A', and we shall define the topological centre $\mathfrak{Z}_t(X')$ of the Banach algebra $(X', \Box)$ which, as a Banach space, is the dual of such a space X.

The next chapter, Chapter 6, is a preliminary to our main work; we shall study the second duals of certain Banach algebras of operators contained in $\mathcal{B}(E)$, the Banach algebra of all bounded linear operators on a Banach space E. A Banach operator algebra contained in $\mathcal{B}(E)$ can only be Arens regular in the case where E is reflexive; we shall sketch a new proof of M. Daws that the Banach algebra $\mathcal{B}(E)$ is indeed Arens regular whenever E is super-reflexive. We shall then discuss the Banach algebras $\mathfrak{A} = \mathcal{K}(E)$ in the case where E is non-reflexive and E' has the approximation property and the Radon–Nikodým property. We shall identify the two topological centres $\mathfrak{Z}_t^{(1)}(\mathfrak{A})$ and $\mathfrak{Z}_t^{(2)}(\mathfrak{A})$ in this case, showing that neither is contained in the other and that each lies strictly between $\mathfrak{A}$ and $\mathfrak{A}''$, and we shall identify their intersection, which may be equal to $\mathfrak{A}$ and which may be strictly larger than $\mathfrak{A}$. We shall also prove that $(\mathfrak{A}'', \Box)$ is semisimple, but that $(\mathfrak{A}'', \Diamond)$ has a 'large' radical. We do not know whether or not $(\mathcal{B}(E)'', \Box)$ is semisimple for all sufficiently 'nice' Banach spaces E; this is true when E is a Hilbert space.

In Chapter 7, we shall introduce the Banach algebras that we shall study. The main characters in our story will be the Beurling algebras $L^1(G,\omega)$ and $\ell^1(G,\omega)$, where ω is a weight function on G, but we shall also introduce some of the relatives of these algebras, such as the C^*-algebra $LUC(G, 1/\omega)$; the latter is a left-introverted subspace of $L^1(G,\omega)'$. At the heart of the narrative is the interplay between the structures of the group G and of the Banach algebras that are constructed on G. For example, we shall recognize the measure algebra $M(G,\omega)$ as the multiplier algebra of the Banach algebra $L^1(G,\omega)$, and show how it can be embedded in $(L^1(G,\omega)'', \Box)$ and in $(LUC(G, 1/\omega)', \Box)$. We shall discuss various 'topologically left-invariant' elements in $L^1(G,\omega)''$, and show how they can be used to prove that the radical of $(L^1(G,\omega)'', \Box)$ is usually non–zero. However, we have been forced to leave open one basic question, namely, whether or not the algebras $L^1(G,\omega)$ (or even the algebras $\ell^1(G,\omega)$) are themselves always semisimple.

There is a considerable difference in behaviour between the algebras $L^1(G,\omega)$ (in the case where G is not discrete) and $\ell^1(G,\omega)$. In Chapter 8, we shall study the algebras $\ell^1(G,\omega)$, and in Chapter 9 and Chapter 10 we shall give a variety of examples that exhibit several phenomena that can occur. A basic example is given by the weight ω_α on $\mathbb{Z}$, where

$$\omega_\alpha(n) = (1 + |n|)^\alpha \quad (n \in \mathbb{Z})$$

for $\alpha \geq 0$. It was shown by Craw and Young [CrY] that the Beurling algebra $\ell^1(\mathbb{Z}, \omega_\alpha)$ is Arens regular if and only if $\alpha > 0$. We shall also exhibit in Chapter 9 various examples in which the topological centre of $\ell^1(\mathbb{Z},\omega)''$ lies strictly between $\ell^1(\mathbb{Z},\omega)$ and $\ell^1(\mathbb{Z},\omega)''$. The most important example seems to be Example 9.15: this exhibits a symmetric weight ω on $\mathbb{Z}$ such that ω is increasing on $\mathbb{Z}^+$, such that $\ell^1(\mathbb{Z},\omega)$ is neither Arens regular nor strongly Arens irregular, and such that the radical of $(\ell^1(\mathbb{Z},\omega)'', \Box)$ is a nilpotent ideal of index exactly 3. We do not know whether or not there is a weight ω on $\mathbb{Z}$ such that the algebra $(\ell^1(\mathbb{Z},\omega)'', \Box)$ is semisimple. In Example 9.17, we shall exhibit a symmetric, unbounded weight ω on $\mathbb{Z}$ such that $\ell^1(\mathbb{Z},\omega)$ is strongly Arens irregular; it is conceivable that for this example $(\ell^1(\mathbb{Z},\omega)'', \Box)$ is indeed semisimple.

In Chapter 10, we shall turn to Beurling algebras on $\mathbb{F}_2$, the free group on two generators. In Theorem 10.12, we shall show by a rather complicated example that the two topological centres of the second dual of a Beurling algebra $\ell^1(\mathbb{F}_2,\omega)$ may be different, and, remarkably, that the radicals of the two second duals may have different indices of nilpotence.

In Chapter 11, we shall study the topological centre $\mathfrak{Z}_t(\mathcal{X}'_\omega)$, where we define $\mathcal{X}_\omega$ to be the space $LUC(G,1/\omega)$, a left-introverted subspace of $L^\infty(G,1/\omega)$. We shall show that $\mathfrak{Z}_t(\mathcal{X}'_\omega)$ can be identified with an algebra $M(G,\omega)$ of measures on G under the condition that ω be diagonally bounded (see Definition 7.41) on a dispersed subset of G. This chapter extends earlier work of Lau and Ülger in [L3] and [LU]; see also [LLo1].

Next, in Chapter 12, we shall turn to a study of the second duals of the algebras $L^1(G,\omega)$ in the case where G is not discrete. We shall show in Theorem 12.2 that $L^1(G,\omega)$ is left strongly Arens irregular whenever $\mathfrak{Z}_t(\mathcal{X}'_\omega) = M(G,\omega)$ (subject to a very mild condition), and hence deduce that the algebra $L^1(G,\omega)$ is strongly Arens irregular whenever ω is diagonally bounded on a dispersed subset of G. Now let

$$\omega_\alpha(t) = (1+|t|)^\alpha \quad (t \in \mathbb{R})$$

for $\alpha \geq 0$, so that ω_α is a weight function on $\mathbb{R}$, but ω_α is not diagonally bounded on any dispersed subset of $\mathbb{R}$ whenever $\alpha > 0$. In an important new result, we shall show in Theorem 12.6 that the Beurling algebra $L^1(\mathbb{R},\omega_\alpha)$ is neither Arens regular nor strongly Arens irregular in the case where $\alpha > 0$, and we shall obtain in Theorem 12.9 and later results rather a large amount of information concerning the radical of the second duals of these algebras.

Finally in Chapter 13, we shall discuss continuous derivations from $L^1(G,\omega)$ into the module $L^1(G,\omega)''$ which is its second dual. We restrict ourselves to the case where the group G is abelian. In particular, we shall determine when many of these algebras are 2-weakly amenable; in our case, this means that each such derivation is 0.

The work concludes with a list of open problems.

We conclude these preliminary remarks by giving some notation that we shall use frequently.

Throughout, we write $\mathbb{N}$ for the set $\{1, 2, \dots\}$ of natural numbers, $\mathbb{Z}^+$ for $\mathbb{N} \cup \{0\}$, $\mathbb{N}_k$ for the set $\{1, 2, \dots, k\}$, and $\mathbb{Z}_k^+$ for the set $\{0, 1, 2, \dots, k\}$. The unit interval $[0, 1]$ is denoted by $\mathbb{I}$, and the unit circle by $\mathbb{T}$. For $z \in \mathbb{C}$ and $r \geq 0$, we set

$$\mathbb{D}(z; r) = \{w \in \mathbb{C} : |w - z| < r\},$$

the open disc with centre z and radius r, and we set $\mathbb{D} = \mathbb{D}(0; 1)$.

The algebra of $n \times n$ matrices over $\mathbb{C}$ is denoted by $\mathbb{M}_n(\mathbb{C})$ or $\mathbb{M}_n$.

For a function f on a set S, the *support* of f is

$$\operatorname{supp} f = \{s \in S : f(s) \neq 0\} .$$

The *characteristic function* of a subset T of S is denoted by χ_T, so that $\chi_T(s) = 1$ whenever $s \in T$ and $\chi_T(s) = 0$ whenever $s \in S \setminus T$.

For subsets S and T of a group G, we set

$$S \cdot T = \{st : s \in S, t \in T\},$$

and $S + T = \{s + t : s \in S, t \in T\}$ in the case where G is abelian and is written additively; we write $S + t$ for $S + \{t\}$, etc. Also, in the general case, we set $S^{-1} = \{s^{-1} : s \in S\}$. An identity of a semigroup S is usually denoted by e_S, and $S^\bullet = S \setminus \{e_S\}$.

Let X be a locally compact space. (By our convention, each locally compact space is taken to be Hausdorff.) Let (s_α) be a net in X. Then

$$\operatorname*{Lim}_{\alpha} s_\alpha = \infty$$

means that, for each compact subset K of X, there exists α_K such that $s_\alpha \in X \setminus K$ $(\alpha \succeq \alpha_K)$. Let (K_α) be a net of compact subsets of X. Then

$$\operatorname*{Lim}_{\alpha} K_\alpha = \infty$$

means that, for each compact subset K of X, there exists α_K such that $K_\alpha \cap K = \emptyset$ $(\alpha \succeq \alpha_K)$. Let $f : X \to \mathbb{C}$ be a function on X. Then

$$\operatorname*{Lim}_{x \to \infty} f(x) = \alpha \quad \text{(respectively,} \quad \operatorname*{Lim\,sup}_{x \to \infty} f(x) = \alpha)$$

means that, for each $\varepsilon > 0$, there is a compact subset K of X such that $|f(x) - \alpha| < \varepsilon$ (respectively, $f(x) < \alpha + \varepsilon$) whenever $x \in X \setminus K$. In particular, we shall use this notation in the case where X is a discrete space.

An index of terms used is given on pages 185–187, and an index of symbols is on pages 189–191.

A few of our results, which we indicate, are based on earlier theorems contained in the PhD thesis [La] at Leeds of David Lamb; we are grateful for his permission to include them here. We are also grateful to Colin Graham and to Matthias Neufang for some valuable comments on and interest in our work and for making their preprints available to us.

This manuscript was completed in May, 2003; a few comments about more recent results, and some extra references, have been added in August, 2004, after the memoir was accepted for publication. Further results in this area will be contained in a memoir of H. G. Dales, A. T.-M. Lau, and D. Strauss, *Banach algebras on compactifications of semigroups*, which is in preparation.

CHAPTER 2

Definitions and Preliminary Results

We begin by recalling some basic concepts and notations. Further details of everything mentioned here are contained in the monograph [D].

Let S be a subset of a linear space. Then $\operatorname{lin} S$ denotes the linear span of S, $\operatorname{ex} S$ denotes the set of extreme points of S, $\langle S\rangle$ is the convex hull of S, and $\operatorname{ac} S$ is the absolutely convex hull of S.

The space of linear maps from a linear space E to a linear space F is denoted by $\mathcal{L}(E,F)$; we write $\mathcal{L}(E)$ for the unital algebra $\mathcal{L}(E,E)$.

Let E be a linear space. Then the image of $x \in E$ under a linear functional λ on E is denoted by $\lambda(x)$ or, more usually, by $\langle x,\, \lambda\rangle$. The space of linear functionals on a linear space E is denoted by $E^{\times}$.

Let A be a (linear, associative, complex) algebra. The product in A is the bilinear map

$$m_A : (a,b) \mapsto ab, \quad A \times A \to A\,.$$

For each $a \in A$, we define

$$L_a(b) = ab, \quad R_a(b) = ba \quad (b \in A)\,;$$

these are the operations of *left* and *right multiplication* by a.

We denote by $A^{\#}$ the algebra formed by adjoining an identity to A (so that $A^{\#} = A$ in the case where A is unital), and by A^{op} the opposite algebra to A, so that A^{op} is the same linear space as A, but the product is $\cdot$, where $a \cdot b = ba$ $(a,b \in A)$.

An element $a \in A$ is *nilpotent* if there exists $n \in \mathbb{N}$ with $a^n = 0$. For each $n \in \mathbb{N}$ and $S \subset A$, we set

$$S^{[n]} = \{a_1 \cdots a_n : a_1, \ldots, a_n \in A\} \quad \text{and} \quad S^n = \operatorname{lin} S^{[n]}\,.$$

For $S, T \subset A$, we set

$$S \cdot T = \{ab : a \in S,\, b \in T\} \quad \text{and} \quad ST = \operatorname{lin} S \cdot T\,;$$

we write aS for $\{a\}S$ when $a \in A$, etc.

The *centre* of the algebra A is denoted by $\mathfrak{Z}(A)$, so that

$$\mathfrak{Z}(A) = \{a \in A : ab = ba \ (b \in A)\}\,.$$

The (Jacobson) *radical* of A is denoted by $\operatorname{rad} A$; the algebra A is *semisimple* if $\operatorname{rad} A = \{0\}$ and *radical* if $\operatorname{rad} A = A$. Let A be a unital algebra with identity e_A: we denote by $\operatorname{Inv} A$ the set of invertible elements in A, and recall that

$$\begin{aligned}\operatorname{rad} A &= \{a \in A : e_A - ba \in \operatorname{Inv} A \ \ (b \in A)\} \\ &= \{a \in A : e_A - ab \in \operatorname{Inv} A \ \ (b \in A)\}\,.\end{aligned}$$

Note that $\operatorname{rad} A = \operatorname{rad} A^{\mathrm{op}}$ as subsets of A.

Let I be a left (respectively, right) ideal in an algebra A. Then I is *left-annihilator* (respectively, *right-annihilator*) if $ax = 0$ (respectively, $xa = 0$) whenever $a \in A$ and $x \in I$; each such ideal is contained in $\operatorname{rad} A$. Let $S \subset A$. For $n \geq 2$, the set S is *nilpotent of index* n if $S^n = \{0\}$, but $S^{n-1} \neq \{0\}$. The radical $\operatorname{rad} A$ contains each left or right ideal which is nilpotent.

PROPOSITION 2.1. *Let I be an ideal in an algebra A, and let $a \in I$. Suppose that $aI = 0$. Then $a \in \operatorname{rad} A$.*

PROOF. Set $J = aA^{\#}$, a right ideal in A. Then

$$J^2 = aA^{\#}aA^{\#} \subset aI = 0\,,$$

and so J is nilpotent. Thus $a \in J \subset \operatorname{rad} A$. □

An algebra A is a *semidirect product* of a subalgebra B and an ideal I if A is the direct sum of B and I as linear spaces; in this case, the product in A is determined by the formula

$$(b_1, x_1)(b_2, x_2) = (b_1 b_2, x_1 b_2 + b_1 x_2 + x_1 x_2) \tag{2.1}$$

for $b_1, b_2 \in B$ and $x_1, x_2 \in I$, and we write $A = B \ltimes I$. The algebra A is *decomposable* if there is a subalgebra B of A such that $A = B \ltimes \operatorname{rad} A$.

Let e be a right identity of an algebra A, so that $ae = a$ $(a \in A)$. Then L_e is a linear projection on A, and L_e is a homomorphism. Define $eA = \{ea : a \in A\}$ and $(1-e)A = \{a - ea : a \in A\}$, so that we have $eA = L_e(A)$ and $(1-e)A = \ker L_e$. Then $A = eA \ltimes (1-e)A$ is a semidirect product.

Let A be an algebra. Then: an element $L \in \mathcal{L}(A)$ such that

$$L(ab) = L(a)b \quad (a, b \in A)$$

is a *left multiplier* on A; an element $R \in \mathcal{L}(A)$ such that

$$R(ab) = aR(b) \quad (a, b \in A)$$

is a *right multiplier* on A; a pair (L, R) such that L is a left multiplier, R is a right multiplier, and

$$aL(b) = R(a)b \quad (a, b \in A)$$

is a *multiplier* on A. For example, for each $a \in A$, L_a is a left multiplier of A, R_a is a right multiplier of A, and (L_a, R_a) is a multiplier of A. The subalgebra of $\mathcal{L}(A) \times \mathcal{L}(A)^{\mathrm{op}}$ consisting of the multipliers (L, R) of A is the *multiplier algebra* of A, denoted by $\mathcal{M}(A)$; the map

$$a \mapsto (L_a, R_a), \quad A \to \mathcal{M}(A),$$

is a homomorphism. In the case where

$$\{a \in A : aA = \{0\}\} = \{a \in A : Aa = \{0\}\} = \{0\}, \tag{2.2}$$

this map is an embedding, and we regard A as a subalgebra of $\mathcal{M}(A)$ in this way. Suppose that A is commutative. Then the multiplier algebra $\mathcal{M}(A)$ is a commutative, unital subalgebra of $\mathcal{L}(A)$ [D, p. 60].

A *linear involution* on a linear space E is a map $x \mapsto x^*$ on E such that

$$\begin{aligned} (x^*)^* &= x \quad (x \in E), \\ (\alpha x + \beta y)^* &= \overline{\alpha} x^* + \overline{\beta} y^* \quad (x, y \in E,\ \alpha, \beta \in \mathbb{C}). \end{aligned}$$

An *involution* on an algebra A is a linear involution on A such that

$$(ab)^* = b^* a^* \quad (a, b \in A).$$

An algebra with an involution is a $*$-*algebra.*

A linear functional λ on a $*$-algebra A is *positive* if

$$\langle aa^*, \lambda \rangle \geq 0 \quad (a \in A);$$

the set $*\text{-rad}\, A$ is defined to be the intersection of the kernels of the positive linear functionals on $A^{\#}$, and the algbera A is $*$-*semisimple* if $*\text{-rad}\, A = \{0\}$. Thus $*\text{-rad}\, A$ is an ideal in A. Let A be a unital algebra with an involution. Then a positive functional λ on A is a *state* if $\langle e_A, \lambda \rangle = 1$; the set of states is the *state space* S_A of A.

Let A and B be algebras. Then the linear space $A \otimes B$ is an algebra for a unique product such that

$$(a_1 \otimes b_1)(a_2 \otimes b_2) = a_1 a_2 \otimes b_1 b_2 \quad (a_1, a_2 \in A,\ b_1, b_2 \in B);$$

see [D, Proposition 1.3.11] for details.

Let A be an algebra, and let E be an A-bimodule with respect to the maps

$$(a, x) \mapsto a \cdot x, \quad (a, x) \mapsto x \cdot a, \quad A \times E \to E.$$

Then E is *symmetric* if $a \cdot x = x \cdot a \ \ (a \in A,\ x \in E)$; a symmetric A-bimodule over a commutative algebra A is termed an A-*module.*

Let A be an algebra, and let E be an A-bimodule. Then $E^{\times}$ is also an A-bimodule for the maps $(a, \lambda) \mapsto a \cdot \lambda$ and $(a, \lambda) \mapsto \lambda \cdot a$, where $a \cdot \lambda$ and $\lambda \cdot a$ are defined by the formulae:

$$\langle x, a \cdot \lambda \rangle = \langle x \cdot a, \lambda \rangle, \quad \langle x, \lambda \cdot a \rangle = \langle a \cdot x, \lambda \rangle \tag{2.3}$$

for $a \in A$, $x \in E$, and $\lambda \in E^{\times}$.

Let E be a left A-module. Then we set

$$A \cdot E = \{a \cdot x : a \in A, x \in E\}, \quad AE = \operatorname{lin} A \cdot E\,,$$

with similar definitions for right A-modules.

DEFINITION 2.2. *Let A be an algebra, and let E be an A-bimodule.*

(i) *The bimodule E is* neo-unital *if $A \cdot E = E \cdot A = E$;*

(ii) *A* derivation *from A into E is an element $D \in \mathcal{L}(A, E)$ such that*

$$D(ab) = a \cdot Db + Da \cdot b \quad (a, b \in A)\,.$$

For example, let $x \in E$, and define $\delta_x : A \to E$ by

$$\delta_x(a) = a \cdot x - x \cdot a \quad (a \in A)\,.$$

Then δ_x is a derivation; such derivations are termed *inner derivations*. In the case where $E = A$, we refer to *derivations on A*.

Now suppose that A is a Banach algebra. The *spectrum* of an element $a \in A$ is denoted by $\sigma(a)$, and the *spectral radius* of a is denoted by $\nu(a)$; an element a is *quasi-nilpotent* if $\nu(a) = 0$, and the set of quasi-nilpotent elements of A is denoted by $\mathfrak{Q}(A)$. Certainly $\operatorname{rad} A$ is a closed ideal in A, and $A/\operatorname{rad} A$ is a semisimple Banach algebra; we have $\operatorname{rad} A \subset \mathfrak{Q}(A)$, and $\operatorname{rad} A = \mathfrak{Q}(A)$ when A is commutative. For example, let $a \in A$. Then, by the spectral radius formula, $a \in \mathfrak{Q}(A)$ if and only if $\lim_{n\to\infty} \|a^n\|^{1/n} = 0$.

The Banach algebra A is *strongly decomposable* if there is a closed subalgebra B of A such that A has the *strong decomposition*

$$A = B \ltimes \operatorname{rad} A$$

as a semi-direct product. Clearly, the Banach algebra A is strongly decomposable if and only if there is a *splitting homomorphism* for the quotient map $\pi : A \to A/\operatorname{rad} A$; this is a continuous homomorphism $\theta : A/\operatorname{rad} A \to A$ such that $\pi \circ \theta$ is the identity on $A/\operatorname{rad} A$. See [BDL] for the theory of decomposable and strongly decomposable Banach algebras.

A *character* on an algebra A is a homomorphism from A onto $\mathbb{C}$; the set of these characters forms the *character space* Φ_A of A. Clearly,

$\Phi_A \subset A^\times$, and we have

$$\varphi \cdot a = a \cdot \varphi = \varphi(a)\varphi \quad (a \in A,\, \varphi \in \Phi_A)\,. \tag{2.4}$$

Every character on a Banach algebra A is continuous, and the space Φ_A is locally compact in the usual Gel'fand topology.

A net (e_α) in a Banach algebra A is a *bounded left approximate identity* for A if $\sup_\alpha \|e_\alpha\| < \infty$ and if $\lim_\alpha \|a - e_\alpha a\| = 0$ for each $a \in A$; *bounded right approximate identities* are defined similarly. A *bounded approximate identity* for A is a net which is both a bounded left and right approximate identity; a bounded approximate identity is *sequential* if the net is indexed by $\mathbb{N}$. Certainly every Banach algebra with a bounded approximate identity satisfies (2.2).

Let E be a Banach space. The closed ball in E with centre 0 and radius $m > 0$ is denoted by $E_{[m]}$, and the dual space of E, the space of continuous linear functionals on E, is denoted by E'; we represent the duality by the pairing

$$(x, \lambda) \mapsto \langle x, \lambda \rangle, \quad E \times E' \to \mathbb{C}\,.$$

The weak topology on E is denoted by $\sigma(E, E')$ and the weak-$*$ topology on E' is $\sigma(E', E)$. The second dual space of E is denoted by E'', with the pairing

$$(\Lambda, \lambda) \mapsto \langle \Lambda, \lambda \rangle, \quad E'' \times E' \to \mathbb{C}\,,$$

and we denote the canonical embedding of E into E'' by κ_E or κ, so that

$$\langle \kappa_E(x), \lambda \rangle = \langle x, \lambda \rangle \quad (x \in E,\, \lambda \in E')\,.$$

Usually we shall regard E as a closed subspace of E'' by identifying E with $\kappa(E)$. The space E is *reflexive* if $\kappa(E) = E''$. We continue to define $E''', E'''', \dots$.

We write $\sigma(E'', E')$ for the weak-$*$ topology on E'', so that $(E'')_{[1]}$ is $\sigma(E'', E')$-compact and $\kappa(E_{[1]})$ is $\sigma(E'', E')$-dense in $(E'')_{[1]}$. Let F be a closed subspace of a Banach space E. Then we identify the space F'' with the $\sigma(E'', E')$-closure of $\kappa_E(F)$ in E''; the relative $\sigma(E'', E')$-topology on F'' coincides with $\sigma(F'', F')$. Note that $E \cap F'' = F$ with these identifications.

Let E be a Banach space. The map $P : E''' \to E'$ which is the dual of the embedding κ of E into E'' (so that $P(\Lambda) = \Lambda \mid \kappa(E)$ for $\Lambda \in E'''$) is a continuous linear projection onto E', called the *canonical projection.*

Let E be a Banach space, and let F and G be closed linear subspaces of E and E', respectively. Then we define the *annihilators*:

$$\begin{aligned} F^\circ &= \{\lambda \in E' : \langle x, \lambda \rangle = 0 \quad (x \in F)\}\,; \\ {}^\circ G &= \{x \in E : \langle x, \lambda \rangle = 0 \quad (\lambda \in G)\}\,. \end{aligned}$$

Clearly F° and ${}^\circ G$ are closed linear subspaces of E' and E, respectively, and ${}^\circ(F^\circ) = F$.

Let E and F be Banach spaces. Then the Banach space of all bounded linear operators from E to F is denoted by $\mathcal{B}(E, F)$; here

$$\|T\| = \sup\{\|Tx\| : x \in E_{[1]}\}$$

defines the operator norm $\|\cdot\|$ on $\mathcal{B}(E, F)$. We write $\mathcal{B}(E)$ for $\mathcal{B}(E, F)$, so that $\mathcal{B}(E)$ is a unital Banach algebra. An operator $T \in \mathcal{B}(E, F)$ is *compact* if $T(E_{[1]})$ is relatively compact in F and *weakly compact* if $T(E_{[1]})$ is relatively weakly compact in F. The spaces of compact and weakly compact operators are denoted by $\mathcal{K}(E, F)$ and $\mathcal{W}(E, F)$, respectively; each is a closed subspace of $\mathcal{B}(E, F)$. We write $\mathcal{K}(E)$ and $\mathcal{W}(E)$ for $\mathcal{K}(E, E)$ and $\mathcal{W}(E, E)$, respectively. Then $\mathcal{K}(E)$ and $\mathcal{W}(E)$ are closed ideals in $\mathcal{B}(E)$. See Chapter 6 for further details.

Again let E and F be Banach spaces. Then the *projective tensor product* of E and F is denoted by $(E \widehat{\otimes} F, \|\cdot\|_\pi)$. Thus each $z \in E \widehat{\otimes} F$ has a representation

$$z = \sum_{j=1}^{\infty} x_j \otimes y_j\,,$$

where $x_j \in E$ and $y_j \in F$ for each $j \in \mathbb{N}$ and

$$\sum_{j=1}^{\infty} \|x_j\|\,\|y_j\| < \infty\,;$$

further, $\|z\|_\pi$ is equal to the infimum of $\sum_{j=1}^{\infty} \|x_j\|\,\|y_j\|$ over all such representations. The dual of $(E \widehat{\otimes} F, \|\cdot\|_\pi)$ is identified with $\mathcal{B}(E, F')$ [D, Proposition A.3.70]. Let A and B be Banach algebras. Then $A \widehat{\otimes} B$ is also a Banach algebra [D, Theorem 2.1.22].

In the case where A is a Banach algebra satisfying (2.2), each left multiplier and each right multiplier on A is a continuous linear operator, $\mathcal{M}(A)$ is a closed subalgebra of the Banach algebra $\mathcal{B}(A) \times \mathcal{B}(A)^{\mathrm{op}}$, and the embedding of A in $\mathcal{M}(A)$ is continuous. However, A is not necessarily a closed subalgebra of $\mathcal{M}(A)$. For a more general theory of multiplier algebras, see [D, Theorem 2.5.12] and [Pa2, Chapter 1.2]; in the latter text, multipliers are termed 'centralisers' of the algebra.

Let A be a Banach algebra satisfying (2.2), and let L and R be left and right multipliers on A. Then we note that

$$(2.5)\quad L'(a \cdot \lambda) = a \cdot L'(\lambda), \quad R'(\lambda \cdot a) = R\lambda \cdot a \quad (a \in A,\, \lambda \in A').$$

A *Banach $*$-algebra* which is a $*$-algebra is a Banach algebra with an isometric involution, as in [D, Chapter 3.1]. In this case, $*\text{-rad}\, A$ is a closed ideal in A, and $*\text{-rad}\, A \supset \operatorname{rad} A$.

A *C^*-algebra* is a Banach $*$-algebra A such that

$$\|aa^*\| = \|a\|^2 \quad (a \in A).$$

Thus a C^*-algebra is a Banach $*$-algebra. A C^*-algebra which, as a Banach space, is the dual of another Banach space is a *von Neumann algebra.* We shall use some standard facts about C^*-algebras and von Neumann algebras; this background material is contained in many texts, including that of Kadison and Ringrose [KR]. Let A be a unital C^*-algebra. Then the state space S_A of A is now equal to

$$\{\lambda \in A' : \|\lambda\| = \langle e_A, \lambda\rangle = 1\}$$

and $\operatorname{lin} S_A = A'$. By the Krein–Milman theorem, $S_A = \overline{\langle \operatorname{ex} S_A \rangle}$, and also $(A')_{[1]} = \overline{\operatorname{ac}}\{\operatorname{ex} S_A\}$, where the closures are taken in the space $(A', \sigma(A', A))$. The elements of $\operatorname{ex} S_A$ are the *pure states.*

Let Ω be a non-empty locally compact space. Then $C_0(\Omega)$ is the space of all complex-valued, continuous functions on Ω such that the functions vanish at infinity; $C_0(\Omega)$ is a commutative C^*-algebra for the pointwise product and the uniform norm, $|\cdot|_\Omega$. We write $C(\Omega)$ in the case where Ω is compact. Further, $C_{00}(\Omega)$ is the dense subalgebra of $C_0(\Omega)$ consisting of all functions whose support is contained in a compact subset of Ω. Each commutative C^*-algebra A is isometrically $*$-isomorphic to $C_0(\Phi_A)$, and $\operatorname{ex} S_A$ is identified with the character space Φ_A; see [KR, 3.4.7] for details. The dual Banach space of $C_0(\Omega)$ is identified with $M(\Omega)$, the space of all complex-valued, regular Borel measures on Ω, with the duality

$$(f, \mu) \mapsto \langle f, \mu\rangle = \int_\Omega f(s)\, \mathrm{d}\mu(s), \quad C_0(\Omega) \times M(\Omega) \to \mathbb{C}.$$

Here $\|\mu\| = |\mu|(\Omega)$ for $\mu \in M(\Omega)$, where $|\mu|$ denotes the total variation of the measure μ.

We denote by $CB(\Omega)$ the space of all bounded, continuous functions on a non-empty, locally compact space Ω, so that $(CB(\Omega), |\cdot|_\Omega)$ is also a commutative C^*-algebra for the pointwise product and the uniform norm. The character space of $CB(\Omega)$ is identified with $\beta\Omega$, the Stone-Čech compactification of Ω, and $CB(\Omega)$ is identified with $C(\beta\Omega)$. For details, see [D, Chapter 4.2].

Let A be a Banach algebra. A *Banach A-bimodule* is a Banach space E such that E is an A-bimodule and

$$\|a \cdot x\| \leq \|a\| \|x\|, \quad \|x \cdot a\| \leq \|a\| \|x\| \quad (a \in A, x \in E).$$

Similarly, we define *Banach left A-modules* and *Banach right A-modules*. For example, A itself is a Banach A-bimodule with respect to the product in A. For each Banach A-bimodule E, the spaces $\overline{AE}$ and $\overline{EA}$ are also Banach A-bimodules; the bimodule E is *essential* if

$$\overline{AE} = \overline{EA} = E.$$

Let A be a Banach algebra with a bounded approximate identity, and let E be an essential Banach A-bimodule. Then E becomes a unital Banach $\mathcal{M}(A)$-bimodule in a natural way. The module operations satisfy the following equations:

$$\left.\begin{array}{rcl} (L, R) \cdot (a \cdot x) & = & La \cdot x, \\ (x \cdot a) \cdot (L, R) & = & x \cdot Ra, \end{array}\right\} \quad (a \in A, (L, R) \in \mathcal{M}(A), x \in E).$$

We shall require the following version of *Cohen's factorization theorem* ([BoDu, Chapter 11], [D, Chapter 2.9], [Pa2, Chapter 5.2]).

THEOREM 2.3. (i) *Let A be a Banach algebra with a bounded right approximate identity. Then $A^{[2]} = A$; if E is a Banach right A-module, then $\overline{EA} = E \cdot A$, and, in particular, $\overline{A'A} = A' \cdot A$.*

(ii) *Let A be a Banach algebra with a bounded approximate identity, and let E be an essential Banach A-bimodule. Then E is neo-unital.* □

In particular, A is neo-unital as a Banach A-bimodule, and so A is a unital Banach $\mathcal{M}(A)$-bimodule. Hence $A', A'', \ldots$ are all $\mathcal{M}(A)$-bimodules; in the case where A is commutative (so that $\mathcal{M}(A)$ is commutative), they are symmetric. For details, see [D, Theorem 2.9.51]; the result is due to B. E. Johnson.

The following result is [LU, Theorem 2.6].

PROPOSITION 2.4. *Let A be a Banach algebra with a sequential bounded approximate identity. Suppose that A is weakly sequentially complete as a Banach space. Then $(A'A)^{\circ} = \{0\}$ if and only if A is unital.* □

Let A be a Banach algebra, and let E be a Banach A-bimodule. Then E' is also a Banach A-bimodule for the maps specified in (2.3). Continuing, we see that E'', E''', ... are also Banach A-bimodules; clearly $\kappa(E)$ is a submodule of E''. In particular, A' is the *dual module* of A, and A'' is the *second dual* module: the canonical embedding $\kappa : A \to A''$ is a module monomorphism and we regard A as a closed submodule of A''. The canonical projection $P : E''' \to E'$ is an A-bimodule homomorphism, and so

$$E''' = E' \oplus \ker P = E' \oplus (\kappa_E(E))^\circ \tag{2.6}$$

as a direct sum of Banach A-bimodules.

The space of continuous derivations from a Banach algebra A into a Banach A-bimodule E is denoted by $\mathcal{Z}^1(A, E)$, and the subspace of inner derivations is $\mathcal{N}^1(A, E)$; we set

$$\mathcal{H}^1(A, E) = \mathcal{Z}^1(A, E)/\mathcal{N}^1(A, E),$$

the *first Banach cohomology group* of A with coefficients in E.

DEFINITION 2.5. *Let A be a Banach algebra. Then A is* amenable *if $\mathcal{H}^1(A, E') = \{0\}$ for each Banach A-bimodule E,* weakly amenable *if $\mathcal{H}^1(A, A') = \{0\}$, and* 2-weakly amenable *if $\mathcal{H}^1(A, A'') = \{0\}$.*

For background information on these concepts (and the more general definition of *n-weakly amenable*), see [BCD], [D], [DGhGr], [He1], and [J1], for example. One of our aims in the present memoir is to determine when certain Beurling algebras are 2-weakly amenable; this will be achieved in Chapter 13.

The following definition is taken from [Ru1, Definition 1.1]; see also [Ru2, Chapter 4.4].

DEFINITION 2.6. *Let A be a Banach algebra. Then A is a* dual *Banach algebra if there is a closed submodule E of the dual module A' such that $E' = A$; the space E is the* predual *of A.*

Let A be a Banach algebra such that $A = E'$ as a Banach space for some Banach space E. Then it is easy to check that A is a dual Banach algebra (with predual E) if and only if the map m_A on $A \times A$ is separately $\sigma(A, E)$-continuous. The predual of a dual Banach algebra is not necessarily unique. (Our notion of a 'dual Banach algebra' is not related to that specified in [BoDu, Definition 32.27].)

For example, a C^*-algebra is a dual Banach algebra if and only if it is a von Neumann algebra [Ru2, Example 4.4.2(c)].

Let A be a Banach algebra. There are two naturally defined products, which we denote by $\Box$ and $\Diamond$, on the Banach space A''. See [D, Chapter 2.7], [DuH], and [Pa2, Chapter 1.4] for a full discussion of these products. We recall the definitions.

For $\lambda \in A'$ and $\Phi \in A''$, define $\lambda \cdot \Phi$ and $\Phi \cdot \lambda$ in A' by the formulae:

$$\langle a, \lambda \cdot \Phi \rangle = \langle \Phi, a \cdot \lambda \rangle, \quad \langle a, \Phi \cdot \lambda \rangle = \langle \Phi, \lambda \cdot a \rangle \quad (a \in A). \tag{2.7}$$

Clearly the maps $(\Phi, \lambda) \mapsto \lambda \cdot \Phi$ and $(\Phi, \lambda) \mapsto \Phi \cdot \lambda$ from $A'' \times A'$ to A' are both bilinear and bounded, with norms equal to 1. We see immediately that $(\lambda \cdot \Phi) \cdot a = \lambda \cdot (\Phi \cdot a)$ and $(a \cdot \lambda) \cdot \Phi = a \cdot (\lambda \cdot \Phi)$, etc., for each $a \in A$, $\lambda \in A'$, and $\Phi \in A''$. Suppose that A is commutative. Then $\lambda \cdot \Phi = \Phi \cdot \lambda$ ($\lambda \in A'$, $\Phi \in A''$).

DEFINITION 2.7. *Let A be a Banach algebra, and let $\Phi, \Psi \in A''$. Then*

$$(2.8)\quad \langle \Phi \,\Box\, \Psi,\, \lambda \rangle = \langle \Phi,\, \Psi \cdot \lambda \rangle, \qquad \langle \Phi \diamond \Psi,\, \lambda \rangle = \langle \Psi,\, \lambda \cdot \Phi \rangle \quad (\lambda \in A')\,.$$

These products $\Box$ and $\diamond$ are the first *and* second Arens products, *respectively, on A''.*

Clearly $\Phi \,\Box\, \Psi, \Phi \diamond \Psi \in A''$ whenever $\Phi, \Psi \in A''$, and both of the maps $(\Phi, \Psi) \mapsto \Phi \,\Box\, \Psi$ and $(\Phi, \Psi) \mapsto \Phi \diamond \Psi$ are bilinear and bounded, with norms equal to 1. Notice also that

$$(2.9)\qquad (\Phi \,\Box\, \Psi) \cdot \lambda = \Phi \cdot (\Psi \cdot \lambda) \quad (\Phi, \Psi \in A'', \lambda \in A')\,.$$

The Arens products are determined by the following formulae, where all limits are taken in the $\sigma(A'', A')$-topology on A''. Let $\Phi, \Psi \in A''$, and take (a_α) and (b_β) to be nets in A such that $a_\alpha \to \Phi$ and $b_\beta \to \Psi$. Then

$$(2.10)\qquad \Phi \,\Box\, \Psi = \lim_\alpha \lim_\beta a_\alpha b_\beta, \qquad \Phi \diamond \Psi = \lim_\beta \lim_\alpha a_\alpha b_\beta\,.$$

It follows from these formulae that both $\Box$ and $\diamond$ are associative products on A''.

THEOREM 2.8. *The algebras $(A'', \Box)$ and $(A'', \diamond)$ are Banach algebras.*

PROOF. This is proved in [D, Chapter 2.6] and [Pa2, Chapter 1.4], for example. $\Box$

It follows easily that the Banach space A' is a Banach left $(A'', \Box)$-module for the map $(\Phi, \lambda) \mapsto \Phi \cdot \lambda$ and a Banach right $(A'', \diamond)$-module for the map $(\Phi, \lambda) \mapsto \lambda \cdot \Phi$.

DEFINITION 2.9. *The Banach algebra A is* Arens regular *if $\Box$ and $\diamond$ coincide on A''.*

We regard A as a closed subalgebra of both $(A'', \Box)$ and $(A'', \diamond)$ by identifying A with $\kappa(A)$. Note that

$$a \,\Box\, \Phi = a \diamond \Phi = a \cdot \Phi \quad \text{and} \quad \Phi \,\Box\, a = \Phi \diamond a = \Phi \cdot a$$

for $a \in A$ and $\Phi \in A''$. Let I be a closed ideal in a Banach algebra A. Then $(I'', \Box)$ is a closed ideal in the Banach algebra $(A'', \Box)$.

It is clear that $(A'', \Diamond) = ((A^{\mathrm{op}})'', \Box)^{\mathrm{op}}$, and so we have

$$\mathrm{rad}\,(A'', \Box) = \mathrm{rad}\,((A^{\mathrm{op}})'', \Diamond)$$

as subsets of A''. In the case where A is commutative, it is immediate that $(A'', \Diamond) = (A'', \Box)^{\mathrm{op}}$, and so A is Arens regular if and only if $(A'', \Box)$ is commutative. Further, for A commutative, we have $\mathrm{rad}\,(A'', \Box) = \mathrm{rad}\,(A'', \Diamond)$ as subsets of A''; each of these radicals is nilpotent of index n if and only if the other has the same property.

PROPOSITION 2.10. *Let A be a Banach algebra with a radical R. Then:*

(i) $(\mathrm{rad}\,(A'', \Box)) \cap A \subset R$; *in the case where A is commutative,*

$$(\mathrm{rad}\,(A'', \Box)) \cap A = R\,;$$

(ii) *A is an ideal in $(A'', \Box)$ if and only if the operators L_a and R_a are both weakly compact for each $a \in A$.*

PROOF. These standard results are given in [D, Proposition 2.6.25] and [Pa2, Proposition 1.4.13], respectively. $\Box$

It seems that no example is known for which $(\mathrm{rad}\,(A'', \Box)) \cap A \neq R$, in the above notation.

For a recent survey of results about Arens regularity, see [FiSi].

Let A be a Banach algebra, and let E be a Banach A-bimodule. Then E'' is a Banach $(A'', \Box)$-bimodule in a natural way (see [DGhGr], [D, Theorem 2.6.15(iii)], [Gou], and [G1]). Briefly, suppose that $\Phi \in A''$ and $\Lambda \in E''$, and take nets (a_α) and (x_β) in A and E, respectively, such that $a_\alpha \to \Phi$ in $(A'', \sigma(A'', A'))$ and $x_\beta \to \Lambda$ in $(E'', \sigma(E'', E'))$. Then

$$\Phi \cdot \Lambda = \lim_\alpha \lim_\beta a_\alpha \cdot x_\beta \quad \text{and} \quad \Lambda \cdot \Phi = \lim_\beta \lim_\alpha x_\beta \cdot a_\alpha$$

in $(E'', \sigma(E'', E'))$.

Suppose that $\theta : A \to B$ is a continuous homomorphism from A into a Banach algebra B. Then B is a Banach A-bimodule for the maps

$$(a, b) \mapsto \theta(a)b \quad \text{and} \quad (a, b) \mapsto b\,\theta(a)\,,$$

and the map $\theta'' : (A'', \Box) \to (B'', \Box)$ is a continuous homomorphism; the module action on B'' described above is given by

$$\Phi \cdot \Lambda = \theta''(\Phi) \,\Box\, \Lambda, \quad \Lambda \cdot \Phi = \Lambda \,\Box\, \theta''(\Phi) \quad (\Phi \in A'', \Lambda \in B'')\,,$$

in this special case. In particular, suppose that A is a closed subalgebra of a Banach algebra B. Then we may regard $(A'', \Box)$ as a closed subalgebra of $(B'', \Box)$.

LEMMA 2.11. *Let A be a Banach algebra, and let E be a Banach A-bimodule. Suppose that $D : A \to E$ is a continuous derivation. Then*

$$D'' : (A'', \Box) \to E''$$

is a continuous derivation.

PROOF. This is [DGhGr, Proposition 1.7]. □

We shall use the following proposition in Chapter 13.

PROPOSITION 2.12. *Let A be a commutative Banach algebra with a bounded approximate identity.*

(i) *The projection $P : A'''' \to A''$ is a $\mathcal{M}(A)$-module homomorphism.*

(ii) *Let $D : A \to A''$ be a continuous derivation. Then there is a continuous derivation $\widetilde{D} : \mathcal{M}(A) \to A''$ such that $\widetilde{D} \mid A = D$.*

PROOF. (i) Take $T \in \mathcal{M}(A)$, $\lambda \in A'$, and $\Phi \in A''''$. Then

$$\begin{aligned} \langle P(T \cdot \Phi), \lambda\rangle &= \langle T \cdot \Phi, \kappa_{A'}(\lambda)\rangle = \langle \Phi, \kappa_{A'}(\lambda) \cdot T\rangle \\ &= \langle \Phi, \kappa_{A'}(\lambda \cdot T)\rangle = \langle P(\Phi), \lambda \cdot T\rangle = \langle T \cdot P(\Phi), \lambda\rangle\,, \end{aligned}$$

and so $P(T \cdot \Phi) = T \cdot P(\Phi)$, as required.

(ii) By Lemma 2.11, $D'' : (A'', \Box) \to A''''$ is a continuous derivation. Set $\widetilde{D} = (P \circ D'') \mid \mathcal{M}(A)$, so that $\widetilde{D} : \mathcal{M}(A) \to A''$ is a continuous linear operator. By (i), $\widetilde{D}$ is a derivation, and clearly $\widetilde{D} \mid A = D$. □

Let A be a Banach $*$-algebra. Then the involution $*$ on A extends to a linear involution $*$ on A'', and

$$(\Phi \,\Box\, \Psi)^* = \Psi^* \diamond \Phi^* \quad (\Phi, \Psi \in A'')\,. \tag{2.11}$$

Indeed, for $\lambda \in A'$, define $\lambda^{\triangleleft} \in A'$ by

$$\langle a, \lambda^{\triangleleft}\rangle = \overline{\langle a^*, \lambda\rangle} \quad (a \in A)\,,$$

and then, for $\Phi \in A''$, define $\Phi^* \in A''$ by

$$\langle \Phi^*, \lambda\rangle = \overline{\langle \Phi, \lambda^{\triangleleft}\rangle} \quad (\lambda \in A')\,. \tag{2.12}$$

It is now easy to see that (2.11) holds. We note also that

$$(\Phi \cdot \lambda)^{\triangleleft} = \lambda^{\triangleleft} \cdot \Phi \quad (\lambda \in A', \Phi \in A'')\,. \tag{2.13}$$

It follows from equation (2.11) that the map $*$ is an involution on $(A'', \Box)$ if and only if A is Arens regular. Let $\Phi \in A''$. Then $\Phi \in \operatorname{rad}(A'', \Box)$ if and only if $\Phi^* \in \operatorname{rad}(A'', \diamond)$; again, each of these radicals is nilpotent of index n if and only if the other has the same property. In the case where A is Arens regular, rad $(A'', \Box)$ is a $*$-ideal in $(A'', \Box)$.

Let A be a Banach algebra, and let $\varphi \in \Phi_A$. Set $\widetilde{\varphi} = \kappa_{A'}(\varphi) \in A'''$. Then it is easily checked that $\widetilde{\varphi}$ is a character on both $(A'', \Box)$ and $(A'', \Diamond)$, and that the map $\varphi \mapsto \widetilde{\varphi}$ is an injection. However this map is usually not continuous. In the case where lin Φ_A is weak-$*$ dense in A' (which holds when A is a commutative C^*-algebra, for example), the image $\kappa_{A'}(\Phi_A)$ is dense in $\Phi_{(A'', \Box)}$.

DEFINITION 2.13. *Let A be a Banach algebra. An element $\Phi_0 \in A''$ is a* mixed identity *for A'' if it is a right identity for $(A'', \Box)$ and a left identity for $(A'', \Diamond)$, so that*

$$\Phi \Box \Phi_0 = \Phi_0 \Diamond \Phi = \Phi \quad (\Phi \in A'') . \tag{2.14}$$

An element $\Phi_0 \in A''$ is a mixed identity if and only if

$$\Phi_0 \cdot \lambda = \lambda \cdot \Phi_0 = \lambda \quad (\lambda \in A') . \tag{2.15}$$

A mixed identity is not necessarily unique for a general Banach algebra, but, in the case where A is Arens regular, a mixed identity is the unique identity of $(A'', \Box)$. Now suppose that A has a bounded approximate identity (e_α). Then (e_α) has a $\sigma(A'', A')$-accumulation point, say Φ_0, in A'', and Φ_0 is a mixed identity for A''. Conversely, if A'' has a mixed identity Φ_0, then A has a bounded approximate identity which converges to Φ_0 in A''. (See [BoDu, p. 146], [CiY], [D, Proposition 2.9.16], or [Pa2, Proposition 5.1.8(a)] for more precise results.)

The following result is given as [D, Theorem 2.9.49].

THEOREM 2.14. *Let A be a Banach algebra such that A'' has a mixed identity Φ_0 with $\|\Phi_0\| = 1$. Then $\mathcal{M}(A)$ is a closed, unital subalgebra of the Banach algebra $\mathcal{B}(A) \times \mathcal{B}(A)^{\mathrm{op}}$, A is isometrically isomorphic to a closed ideal in $\mathcal{M}(A)$, and the map*

$$\kappa : (L, R) \mapsto \Phi_0 \cdot (L, R) = R''(\Phi_0), \quad \mathcal{M}(A) \to (A'', \Box), \tag{2.16}$$

is an isometric embedding such that $\kappa \mid A = \kappa_A$. Further,

$$\mathcal{M}(A) \subset \Phi_0 \Box A'' .$$

$\Box$

Let A be a dual Banach algebra, with predual space E, so that

$$E^\circ = \{\Phi \in A'' : \Phi \mid E = 0\} .$$

Then the canonical projection $P : E''' \to E'$ gives a continuous linear map $P : A'' \to A$, and now equation (2.6) can be written as

$$A'' = A \oplus E^\circ$$

as a direct sum of Banach A-bimodules. For each $\Phi \in A''$, there exists $a = P(\Phi) \in A$ such that $\langle \Phi, \lambda \rangle = \langle a, \lambda \rangle$ $(\lambda \in E)$, and clearly we have $\Phi \cdot \lambda = a \cdot \lambda$ $(\lambda \in E)$. It follows from (2.8) that

$$P(\Phi \,\square\, \Psi) = P(\Phi) \cdot P(\Psi) \quad (\Phi, \Psi \in A''),$$

and so $P : (A'', \square\,) \to (A, \cdot)$ is a homomorphism and E° is a closed ideal in $(A'', \square\,)$. (See also [GhLaa, Theorem 2.2].) Consider the short exact sequence

$$\textstyle\sum : 0 \longrightarrow E^\circ = \ker P \longrightarrow (A'', \square\,) \xrightarrow{P} (A, \cdot\,) \longrightarrow 0 \tag{2.17}$$

of Banach algebras and continuous homomorphisms. Clearly $P \circ \kappa_A$ is the identity map on A, and so κ_A is a splitting homomorphism for $\sum$. Thus

$$(A'', \square\,) = A \ltimes E^\circ \tag{2.18}$$

as a semidirect product. Similarly, the map $P : (A'', \Diamond\,) \to (A, \cdot\,)$ is a homomorphism, E° is a closed ideal in $(A'', \Diamond\,)$, and there is a strong decomposition $(A'', \Diamond\,) = A \ltimes E^\circ$. Thus we have the following basic theorem.

THEOREM 2.15. *Let A be a dual Banach algebra with predual space E. Then we have $(A'', \square\,) = A \ltimes E^\circ$ as a semidirect product. In the case where A is semisimple,*

$$\operatorname{rad}(A'', \square\,) \subset E^\circ. \tag{2.19}$$

□

In general, $\operatorname{rad}(A'', \square\,) \neq E^\circ$. For example, in the case where A is a C^*-algebra, as in Example 4.2, below, A is Arens regular and $(A'', \square\,)$ is semisimple, and indeed $(A'', \square\,)$ is a C^*-algebra. In particular, for a von Neumann algebra A with predual E, we have $\operatorname{rad}(A'', \square\,) \neq E^\circ$, and it is certainly not true that $\Phi \,\square\, \Psi = 0$ $(\Phi, \Psi \in E^\circ)$. However, we do have the following results.

PROPOSITION 2.16. *Let A be a dual Banach algebra, with predual space E.*

(i) *Suppose that $\Phi \,\square\, \Psi = \Phi \Diamond \Psi = 0$ $(\Phi, \Psi \in E^\circ)$. Then*

$$(a, \Phi) \,\square\, (b, \Psi) = (a, \Phi) \Diamond (b, \Psi) = (ab, a \cdot \Psi + \Phi \cdot b) \tag{2.20}$$

for $a, b \in A$ and $\Phi, \Psi \in E^\circ$, and A is Arens regular.

(ii) *Suppose, further, that A is semisimple. Then*

$$\operatorname{rad}(A'', \square\,) = \operatorname{rad}(A'', \Diamond\,) = E^\circ,$$

and $(A'', \square\,) = A \ltimes E^\circ$ is a strong decomposition of $(A'', \square\,)$.

PROOF. (i) This is immediate.

(ii) By hypothesis, the closed ideal E° is nilpotent in $(A'', \Box)$ and $(A'', \Diamond)$, and so $E^\circ \subset \mathrm{rad}(A'', \Box)$ and $E^\circ \subset \mathrm{rad}(A'', \Diamond)$. Since A is semisimple, $\mathrm{rad}(A'', \Box) \subset E^\circ$ and $\mathrm{rad}(A'', \Diamond) \subset E^\circ$. $\Box$

Suppose that A satisfies the following condition:

$$\Phi \cdot \lambda,\ \lambda \cdot \Phi \in E \quad \text{whenever} \quad \Phi \in E^\circ \quad \text{and} \quad \lambda \in A'. \tag{2.21}$$

Take $\Phi, \Psi \in E^\circ$. Then, by the definitions in equations (2.8), we have $\Phi \,\Box\, \Psi = \Phi \Diamond \Psi = 0$, and so the further condition specified in the above proposition is satisfied.

Suppose that A is a commutative Banach algebra and that we have $\Phi \,\Box\, \Psi = 0 \ (\Phi, \Psi \in E^\circ)$. Then we also have $\Phi \Diamond \Psi = 0 \ (\Phi, \Psi \in E^\circ)$. However this implication does not hold for general, non-commutative Banach algebras A; see Theorem 10.12, below.

We now come to a key concept of this memoir.

DEFINITION 2.17. *Let A be a Banach algebra. Then the* topological centres $\mathfrak{Z}_t^{(1)}(A'')$ *and* $\mathfrak{Z}_t^{(2)}(A'')$ *of* A'' *are:*

$$\begin{aligned} \mathfrak{Z}_t^{(1)}(A'') &= \{\Phi \in A'' : \Phi \,\Box\, \Psi = \Phi \Diamond \Psi \ (\Psi \in A'')\}\,; \\ \mathfrak{Z}_t^{(2)}(A'') &= \{\Phi \in A'' : \Psi \,\Box\, \Phi = \Psi \Diamond \Phi \ (\Psi \in A'')\}\,. \end{aligned}$$

Let $\Phi \in \mathfrak{Z}(A'', \Box)$. Then $\lambda \cdot \Phi = \Phi \cdot \lambda \ (\lambda \in A')$, and so, for each $\Psi \in A''$ and $\lambda \in A'$, we have

$$\langle \Phi \,\Box\, \Psi,\ \lambda \rangle = \langle \Psi \,\Box\, \Phi,\ \lambda \rangle = \langle \Psi,\ \Phi \cdot \lambda \rangle = \langle \Psi,\ \lambda \cdot \Phi \rangle = \langle \Phi \Diamond \Psi,\ \lambda \rangle\,,$$

whence $\Phi \,\Box\, \Psi = \Phi \Diamond \Psi$. Thus $\mathfrak{Z}(A'', \Box) \subset \mathfrak{Z}_t^{(1)}(A'')$. Similarly, we have $\mathfrak{Z}(A'', \Box) \subset \mathfrak{Z}_t^{(2)}(A'')$.

It is easy to see that $\mathfrak{Z}_t^{(1)}(A'')$ and $\mathfrak{Z}_t^{(2)}(A'')$ are both $\|\cdot\|$-closed subalgebras of both $(A'', \Box)$ and $(A'', \Diamond)$, and that, for example, $\mathfrak{Z}_t^{(1)}(A'')$ is the set of elements $\Phi \in A''$ such that the map

$$L_\Phi : \Psi \mapsto \Phi \,\Box\, \Psi, \quad A'' \to A'',$$

is continuous when A has the $\sigma(A'', A')$-topology. Clearly

$$A \subset \mathfrak{Z}_t^{(1)}(A'') \subset A'' \quad \text{and} \quad A \subset \mathfrak{Z}_t^{(2)}(A'') \subset A'' \tag{2.22}$$

in each case, and the algebra A is Arens regular if and only if either $\mathfrak{Z}_t^{(1)}(A'') = A''$ or $\mathfrak{Z}_t^{(2)}(A'') = A''$, in which case the two topological centres are both equal to A''. We shall see in Example 6.2 and Theorem 10.12, below, that these two topological centres may be different. However, in the case where A is commutative, $\mathfrak{Z}_t^{(1)}(A'') = \mathfrak{Z}_t^{(2)}(A'') = \mathfrak{Z}(A'')$, the centre of both of the algebras $(A'', \Box)$ and $(A'', \Diamond)$.

Let A be a Banach algebra. Then we have

$$\mathfrak{Z}_t^{(1)}(A'') = \mathfrak{Z}_t^{(2)}((A^{\text{op}})'') \quad \text{and} \quad \mathfrak{Z}_t^{(2)}(A'') = \mathfrak{Z}_t^{(1)}((A^{\text{op}})'');$$

in the case where A is a Banach $*$-algebra, it follows immediately from (2.11) that, for each $\Phi \in A''$, we have $\Phi \in \mathfrak{Z}_t^{(1)}(A'')$ if and only if $\Phi^* \in \mathfrak{Z}_t^{(2)}(A'')$.

The two separate spaces $\mathfrak{Z}_t^{(1)}(A'')$ and $\mathfrak{Z}_t^{(2)}(A'')$ were first considered in the paper [LU], where they are denoted by Z_1 and Z_2, respectively. Some questions raised in [LU] are answered in [GhMMe]; see also [GhLaa].

Let B be a closed subalgebra of A, so that $(B'', \Box)$ and $(B'', \Diamond)$ are $\|\cdot\|$-closed subalgebras of $(A'', \Box)$ and $(A'', \Diamond)$, respectively. Then clearly we have

$$\left(B'' \cap \mathfrak{Z}_t^{(i)}(A'')\right) \subset \mathfrak{Z}_t^{(i)}(B'') \quad (i = 1, 2). \tag{2.23}$$

We shall see in Example 9.3 that the above inclusion can be strict.

Let I be a closed ideal in a Banach algebra A, set $B = A/I$, and take $q : A \to B$ to be the quotient map, so that $q'' : A'' \to B''$ is a continuous surjection with kernel $I^{\circ\circ}$, identified with the weak-$*$ closure of I in A''. Clearly $I^{\circ\circ}$ is a closed ideal in $(A'', \Box)$ and $(A'', \Diamond)$, and $(B'', \Box) = (A'', \Box)/I^{\circ\circ}$ and $(B'', \Diamond) = (A'', \Diamond)/I^{\circ\circ}$. Further,

$$q''\left(\mathfrak{Z}_t^{(i)}(A'')\right) \subset \mathfrak{Z}_t^{(i)}(B'') \quad (i = 1, 2). \tag{2.24}$$

We shall see in Examples 4.3 and 4.6 that this inclusion can be strict.

The following terminology (but not the concept) is new.

DEFINITION 2.18. *Let A be a Banach algebra. Then A is* left (*respectively,* right) strongly Arens irregular *if $\mathfrak{Z}_t^{(1)}(A'') = A$ (respectively, $\mathfrak{Z}_t^{(2)}(A'') = A$), and A is* strongly Arens irregular *if*

$$\mathfrak{Z}_t^{(1)}(A'') = \mathfrak{Z}_t^{(2)}(A'') = A.$$

An easy example of a Banach algebra which is left strongly Arens irregular, but not right strongly Arens irregular, will be given in Example 4.5, below.

Let A be a dual Banach algebra, with predual E. Then A is strongly Arens irregular if and only if, for each $\Phi \in E^\circ \setminus \{0\}$, there exist $\Psi_1 \in E^\circ$ with $\Phi \Box \Psi_1 \neq \Phi \Diamond \Psi_1$ and $\Psi_2 \in E^\circ$ with $\Psi_2 \Box \Phi \neq \Psi_2 \Diamond \Phi$. Clearly A is left strongly Arens irregular if and only if A^{op} is right strongly Arens irregular.

Let A be a commutative Banach algebra. Then A is strongly Arens irregular if and only if $\mathfrak{Z}(A'') = A$.

PROPOSITION 2.19. *Let A be a commutative Banach algebra. Suppose that $A'' = A \ltimes I$, where I is a nilpotent ideal of index $n \geq 2$ in $(A'', \Box)$. Then $I^{n-1} \subset \mathfrak{Z}(A'')$, and A is not strongly Arens irregular.*

PROOF. Let $\Phi \in I^{n-1}$, and take $\Psi \in I$ and $a \in A$. Then

$$\Phi \Box (a + \Psi) = \Phi \cdot a = a \cdot \Phi = (a + \Psi) \Box \Phi ,$$

and so $\Phi \in \mathfrak{Z}(A'')$. Since $I^{n-1} \not\subset A$, the algebra A is not strongly Arens irregular. $\Box$

The following is a typical short calculation involving a centre.

PROPOSITION 2.20. *Let A be a Banach algebra. Then*

$$A' \cdot \mathfrak{Z}_t^{(1)}(A'') \subset \overline{A'A} .$$

PROOF. Take $\lambda \in A'$ and $\Phi \in \mathfrak{Z}_t^{(1)}(A'')$.

Let (a_α) be a net in A with $\lim_\alpha a_\alpha = \Phi$ in the weak-$*$ topology. For each Ψ in A'', we have

$$\begin{aligned} \langle \Psi, \lambda \cdot \Phi \rangle &= \langle \Phi \Diamond \Psi, \lambda \rangle = \langle \Phi \Box \Psi, \lambda \rangle = \langle \Phi, \Psi \cdot \lambda \rangle \\ &= \lim_\alpha \langle a_\alpha, \Psi \cdot \lambda \rangle = \lim_\alpha \langle \Psi, \lambda \cdot a_\alpha \rangle . \end{aligned}$$

Thus $\lambda \cdot \Phi$ belongs to the closure of $A'A$ in the weak topology; by Mazur's theorem (see [D, Theorem A.3.29(ii)]), the latter set is $\overline{A'A}$, the $\|\cdot\|$-closure of $A'A$ in $(A', \|\cdot\|)$. $\Box$

THEOREM 2.21. *Let A be a Banach algebra such that A'' has a mixed identity Φ_0. Then $\mathfrak{Z}_t^{(2)}(A'') \subset \Phi_0 \Box A''$. Suppose, further, that $\overline{A'A} \neq A'$. Then $\Phi_0 \notin \mathfrak{Z}_t^{(1)}(A'')$ and A is not Arens regular.*

PROOF. Let $\Phi \in \mathfrak{Z}_t^{(2)}(A'')$. Then $\Phi = \Phi_0 \Diamond \Phi = \Phi_0 \Box \Phi$. This shows that $\mathfrak{Z}_t^{(2)}(A'') \subset \Phi_0 \Box A''$.

Set $X = \overline{A'A}$. Assume towards a contradiction that $\Phi_0 \in \mathfrak{Z}_t^{(1)}(A'')$, and take $\Psi \in X^\circ$. Then $\Phi_0 \Diamond \Psi = \Phi_0 \Box \Psi$. But $\Phi_0 \Box \Psi = 0$ and $\Phi_0 \Diamond \Psi = \Psi$, and so $\Psi = 0$. This is a contradiction in the case where $X^\circ \neq \{0\}$. $\Box$

The following result was proved by Ülger by an elegant argument in [U4, Theorem 3.3]. The result follows from Proposition 2.4 and Theorem 2.21 in the case where A has a sequential bounded approximate identity, and a short further argument gives the general result; for details, see [D, Theorem 2.9.39].

THEOREM 2.22. *Let A be a non-unital Banach algebra with a bounded approximate identity. Suppose that A is weakly sequentially complete as a Banach space. Then A is not Arens regular.* □

CHAPTER 3

Repeated Limit Conditions

There is a large variety of conditions that determine when a Banach algebra is Arens regular. One of these involves a standard 'repeated limit' condition that descends from a condition of Grothendieck [Gth], and was first utilized in our context by Pym [Py1]. We describe this condition. The results of this chapter are basically known, and can be found in the reference [BJM] and [Y1]; however our notation and formulations are different, and so we give the details.

Let S and T be non-empty sets, let $f : S \times T \to \mathbb{C}$ be a function, and let (s_m) and (t_n) be two sequences in S and T, respectively. We shall usually write

$$\lim_m \lim_n f(s_m, t_n) \quad \text{for} \quad \lim_{m\to\infty}\left(\lim_{n\to\infty} f(s_m, t_n)\right),$$

etc. We say that $\lim_m \lim_n f(s_m, t_n)$ and $\lim_n \lim_m f(s_m, t_n)$ are *repeated limits* of the double sequence $(f(s_m, t_n) : m, n \in \mathbb{N})$. Similarly, we write

$$\operatorname*{Lim}_{x\to\infty} \operatorname*{Lim}_{y\to\infty} f(x, y) \quad \text{for} \quad \operatorname*{Lim}_{x\to\infty}\left(\operatorname*{Lim}_{y\to\infty} f(x, y)\right)$$

for a function $f : X \times Y \to \mathbb{C}$, where X and Y are locally compact spaces.

We record the following well-known triviality, which will be used several times.

PROPOSITION 3.1. *Let S and T be non-empty sets, and let*

$$f : S \times T \to \mathbb{C}$$

be a function. Suppose that (s_α) and (t_β) are nets in S and T, respectively, such that $a = \lim_\alpha \lim_\beta f(s_\alpha, t_\beta)$ and $b = \lim_\beta \lim_\alpha f(s_\alpha, t_\beta)$ both exist. Then there are subsequences (s_{α_m}) and (t_{β_n}) of the nets (s_α) and (t_β), respectively, such that $a = \lim_m \lim_n f(s_{\alpha_m}, t_{\beta_n})$ and $b = \lim_n \lim_m f(s_{\alpha_m}, t_{\beta_n})$.

PROOF. Set $a_\alpha = \lim_\beta f(s_\alpha, t_\beta)$ and $b_\beta = \lim_\alpha f(s_\alpha, t_\beta)$ for each α and β.

We *claim* that there exist subsequences (s_{α_m}) of (s_α) and (t_{β_n}) of (t_β) such that:

$$\begin{array}{rcl} |a - a_{\alpha_m}| < 1/m\,; & & |b - b_{\beta_n}| < 1/n\,; \\ |b_{\beta_n} - f(s_{\alpha_m}, t_{\beta_n})| & < & 1/m \quad (n \in \mathbb{N}_{m-1})\,; \\ |a_{\alpha_m} - f(s_{\alpha_m}, t_{\beta_n})| & < & 1/n \quad (m \in \mathbb{N}_{n-1})\,. \end{array}$$

First choose α_1 and β_1 so that $|a - a_{\alpha_1}| < 1$ and $|b - b_{\beta_1}| < 1$.

Now assume that $\alpha_1, \dots, \alpha_k$ and $\beta_1, \dots, \beta_k$ have been chosen appropriately. Choose α_{k+1} so that

$$\left|a - a_{\alpha_{k+1}}\right| < \frac{1}{k+1}\,, \quad \left|b_{\beta_i} - f(s_{\alpha_{k+1}}, t_{\beta_i})\right| < \frac{1}{k+1} \quad (i \in \mathbb{N}_k)\,,$$

and then choose β_{k+1} so that

$$\left|b - b_{\beta_{k+1}}\right| < \frac{1}{k+1}\,, \quad \left|a_{\alpha_i} - f(s_{\alpha_i}, t_{\beta_{k+1}})\right| < \frac{1}{k+1} \quad (i \in \mathbb{N}_{k+1})\,.$$

This continues the inductive construction of the sequences.

Clearly $\lim_n f(s_{\alpha_m}, t_{\beta_n}) = a_{\alpha_m}$ for each $m \in \mathbb{N}$ and $\lim_m a_{\alpha_m} = a$ and so $a = \lim_m \lim_n f(s_{\alpha_m}, t_{\beta_n})$. Similarly, $b = \lim_n \lim_m f(s_{\alpha_m}, t_{\beta_n})$. □

The following terminology is a modification of that given in [BaR].

DEFINITION 3.2. *Let X and Y be non-empty sets, and let*

$$f : X \times Y \to \mathbb{C}$$

be a function. Then:

(i) f clusters on $X \times Y$ *if*

$$\lim_m \lim_n f(x_m, y_n) = \lim_n \lim_m f(x_m, y_n)$$

whenever (x_m) and (y_n) are sequences in X and Y, respectively, each consisting of distinct points, and both repeated limits exist;

(ii) f 0-clusters on $X \times Y$ *if*

$$\lim_m \lim_n f(x_m, y_n) = \lim_n \lim_m f(x_m, y_n) = 0$$

whenever (x_m) and (y_n) are sequences in X and Y, respectively, each consisting of distinct points, and both repeated limits exist.

Suppose that $h : X \times Y \to \mathbb{C}$ is bounded and that (x_m) and (y_n) are sequences of distinct points in X and Y, respectively. Then we can find subsequences (x_{m_j}) and (y_{n_k}) of (x_m) and (y_n), respectively, such that the two repeated limits of $(h(x_{m_j}, y_{n_k}) : j, k \in \mathbb{N})$ both exist. This shows that h fails to 0-cluster on $X \times Y$ if and only if there exist sequences (x_m) and (y_n) of distinct points in X and Y, respectively, such that one of the two repeated limits of $(h(x_m, y_n) : m, n \in \mathbb{N})$ exists and is non-zero. It also implies that $f+g$ and fg both cluster on $X \times Y$ whenever $f, g : X \times Y \to \mathbb{C}$ are both bounded functions that cluster on $X \times Y$.

There is a slight generalization of the above definition. Let $k \in \mathbb{N}$ with $k \geq 2$, let $X_1, \ldots, X_k$ be non-empty sets, and let

$$f : X_1 \times \cdots \times X_k \to \mathbb{C}$$

be a function. Then f *clusters on* $X_1 \times \cdots \times X_k$ if all k-fold repeated limits of the sequence

$$(f(x_{n_1}, \ldots, x_{n_k}) : (n_1, \ldots, n_k) \in \mathbb{N}^k)$$

are equal whenever all $k!$ such limits exist, and f 0-*clusters on* the set $X_1 \times \cdots \times X_k$ if, further, all these limits are equal to 0. For a definition that is more generally applicable, see [Y1].

The following result of Grothendieck [Gth] is contained in [BJM, Chapter 4, Theorem 2.3], for example. We give a proof (essentially from [Y1, Theorem 1]) for completeness and because we shall require the exact form of the result at some key points.

Let X and Y be non-empty, locally compact sets, and let

$$f : X \times Y \to \mathbb{C}$$

be a bounded, separately continuous function. For $y \in Y$, we set

$$f_y : x \mapsto f(x, y), \quad X \to \mathbb{C},$$

and we regard f_y as an element of $CB(X) = C(\beta X)$; we then set

$$f(x, y) = f_y(x) \quad (x \in \beta X,\ y \in Y).$$

In the next two results, we set $\mathcal{F} = \{f_y : y \in Y\}$.

THEOREM 3.3. *Let X and Y be non-empty, locally compact spaces, and let $f : X \times Y \to \mathbb{C}$ be a bounded, separately continuous function. Then the following conditions are equivalent:*

(a) *$\mathcal{F}$ is relatively weakly compact in $C(\beta X)$;*

(b) *$\mathcal{F}$ is relatively weakly sequentially compact in $C(\beta X)$;*

(c) *$\langle \mathcal{F} \rangle$ is relatively weakly sequentially compact in $C(\beta X)$;*

(d) *f clusters on $X \times Y$;*

(e) *f has an extension to a unique separately continuous function $f : \beta X \times \beta Y \to \mathbb{C}$.*

PROOF. (a) $\Leftrightarrow$ (b) This is the Eberlein–Smulian theorem.

(a) $\Leftrightarrow$ (c) This is the Krein–Smulian theorem.

(a) $\Rightarrow$ (d) Assume that $\mathcal{F}$ is relatively weakly compact in $C(\beta X)$, and take (x_m) and (y_n) to be sequences in X and Y, respectively, such that the two repeated limits of $(f(x_m, y_n) : m, n \in \mathbb{N})$ both exist. Let $h \in C(\beta X)$ be a weak accumulation point of $\{f_{y_n} : n \in \mathbb{N}\}$, and let x be an accumulation point of $\{x_m : m \in \mathbb{N}\}$ in βX. Then

$$\lim_n \lim_m f(x_m, y_n) = \lim_n f_{y_n}(x) = h(x)$$

and

$$\lim_m \lim_n f(x_m, y_n) = \lim_m h(x_m) = h(x)\,,$$

where we are using the fact that various limits exist, and so f clusters on $X \times Y$.

(d) $\Rightarrow$ (b) We shall first show that $\mathcal{F}$ is relatively compact in the pointwise topology of $C(\beta X)$. Choose $k > 0$ such that

$$|f(x, y)| \leq k \quad (x \in X,\, y \in Y)\,,$$

and regard $\mathcal{F}$ as a subset of the compact space $\prod\{D_x : x \in \beta X\}$, where $D_x = \overline{\mathbb{D}(0; k)}$ for each $x \in \beta X$. Take h to be an element of the closure of $\mathcal{F}$ in this space.

We *claim* that h is continuous on βX. For assume towards a contradiction that this is not the case. Then there exists $x_0 \in \beta X$ and $\delta > 0$ such that each neighbourhood of x_0 in βX contains a point $x \in X$ such that $|h(x_0) - h(x)| \geq \delta$. Construct sequences (x_m) in X and (y_n) in Y as follows. First choose any $y_1 \in Y$, and then choose $x_1 \in X$ so that

$$|f(x_0, y_1) - f(x_1, y_1)| < 1 \quad \text{and} \quad |h(x_1) - h(x_0)| \geq \delta\,;$$

this is possible because the function f_{y_1} is continuous on the space X. Having specified $x_1, \ldots, x_n$ in X and $y_1, \ldots, y_n$ in Y, choose $y_{n+1} \in Y$ such that

$$|f(x_i, y_{n+1}) - h(x_i)| < 1/n \quad (i \in \mathbb{Z}_n^+)\,,$$

and then choose $x_{n+1} \in X \setminus \{x_1, \ldots, x_n\}$ such that

$$|f(x_0, y_i) - f(x_{n+1}, y_i)| < 1/n \quad (i \in \mathbb{N}_{n+1})$$

and

$$|h(x_0) - h(x_{n+1})| \geq \delta\,;$$

again, this is possible because the functions $f_{y_1}, \ldots, f_{y_{n+1}}$ are each continuous on the space X. We obtain sequences (x_m) in X and (y_n) in Y. The sequence (x_m) consists of distinct points. Assume that $y_n = y$ for infinitely many $n \in \mathbb{N}$. Then $f(x_m, y) = h(x_m)$ $(m \in \mathbb{N})$ and

$\lim_m f(x_m, y) = h(x_0)$, a contradiction. Hence we may suppose that the sequence (y_n) consists of distinct points.

Clearly

$$\lim_n \lim_m f(x_m, y_n) = \lim_n f(x_0, y_n) = h(x_0)$$

and $\lim_n f(x_m, y_n) = h(x_m)$ $(m \in \mathbb{N})$. By passing to a subsequence of the sequence (x_m), we may suppose that $\lim_m h(x_m) = \alpha$ for some $\alpha \in \overline{\mathbb{D}(0;k)}$; we have $|h(x_0) - \alpha| \geq \delta$. It is now the case that the two repeated limits of the double sequence $(f(x_m, y_n) : m, n \in \mathbb{N})$ both exist (being $h(x)$ and α, respectively), but that these two limits are unequal. Since f clusters on $X \times Y$, this is a contradiction, and so $h \in C(\beta X)$, as claimed.

Now let (f_n) be a sequence in $\mathcal{F}$. Then, by passing to a subsequence, we may suppose that there exists $h \in C(\beta X)$ such that $f_n \to h$ pointwise on βX. By the dominated convergence theorem,

$$\int_{\beta X} f_n \, d\mu \to \int_{\beta X} h \, d\mu$$

for each measure μ on βX, and so $f_n \to h$ weakly. Thus $\mathcal{F}$ is relatively weakly sequentially compact.

(d) $\Leftrightarrow$ (e) This is now clear. □

The following famous condition of Grothendieck is an easy corollary of the above result.

PROPOSITION 3.4. *Let E and F be Banach spaces, and suppose that $T \in \mathcal{B}(E, F)$. Then T is weakly compact if and only if the function*

$$(x, \lambda) \mapsto \langle Tx, \lambda \rangle, \quad E \times F' \to \mathbb{C},$$

clusters on $E_{[1]} \times F'_{[1]}$. □

PROPOSITION 3.5. *Let X and Y be non-empty, locally compact spaces, and let $f : X \times Y \to \mathbb{C}$ be a bounded, separately continuous function that 0-clusters on $X \times Y$. Let $x \in \beta X \setminus X$ and $\varepsilon > 0$. Then there is a finite subset F of Y such that $|f(x, y)| < \varepsilon$ $(y \in Y \setminus F)$.*

PROOF. Assume towards a contradiction that there is no such set F. Then there is a sequence (y_n) of distinct points of Y such that

$$|f(x, y_n)| \geq \varepsilon \quad (n \in \mathbb{N}).$$

Since f clusters on $X \times Y$, it follows from Theorem 3.3 that $\mathcal{F}$ is relatively weakly sequentially compact in $C(\beta X)$, and so, passing to a subsequence of (y_n), there exists $h \in C(\beta X)$ such that $f_{y_n} \to h$ weakly in $C(\beta X)$. Inductively choose a sequence (x_m) of distinct points in X

so that $|f(x_m, y_n)| > \varepsilon/2$ $(n \in \mathbb{N}_m)$ for each $m \in \mathbb{N}$. We may suppose that the two repeated limits of $(f(x_m, y_n) : m, n \in \mathbb{N})$ both exist, say $\alpha = \lim_n \lim_m f(x_m, y_n)$. Then $|\alpha| \geq \varepsilon/2$, a contradiction of the fact that f 0-clusters on $X \times Y$.

This establishes the result. □

The following result is similar to Theorem 3.3; see [Y1, Corollary 1 to Theorem 2].

PROPOSITION 3.6. *Let $k \in \mathbb{N}$ with $k \geq 2$, let $X_1, \dots, X_k$ be non-empty, locally compact spaces, and let $f : X_1 \times \cdots \times X_k \to \mathbb{C}$ be a bounded, separately continuous function. Then the function f has an extension to a separately continuous function $f : \beta X_1 \times \cdots \times \beta X_k \to \mathbb{C}$ if and only if f clusters on $X_1 \times \cdots \times X_k$. Further, in this case,*

$$f \mid (\beta X_1 \setminus X_1) \times \cdots \times (\beta X_k \setminus X_k) = 0$$

if and only if f 0-clusters on $X_1 \times \cdots \times X_k$. □

The concepts defined in the next definition will be required in Chapter 12; although more general definitions could be given, we restrict ourselves to bounded, continuous functions f.

DEFINITION 3.7. *Let X and Y be non-empty, locally compact spaces, and let $f : X \times Y \to \mathbb{C}$ be a bounded, continuous function. Then:*

(i) f 0-clusters strongly on $X \times Y$ *if*

$$\operatorname*{Lim}_{x\to\infty} \operatorname*{Lim\,sup}_{y\to\infty} f(x,y) = \operatorname*{Lim}_{y\to\infty} \operatorname*{Lim\,sup}_{x\to\infty} f(x,y) = 0\,;$$

(ii) f 0-clusters locally uniformly on $X \times Y$ *if*

$$\lim_{\alpha} \limsup_{\beta} \sup_{K_\alpha \times L_\beta} f = \lim_{\beta} \limsup_{\alpha} \sup_{K_\alpha \times L_\beta} f = 0$$

whenever (K_α) and (L_β) are nets of compact subsets of X and Y, respectively, such that $\operatorname{Lim}_\alpha K_\alpha = \operatorname{Lim}_\beta L_\beta = \infty$.

Let $f : \mathbb{N} \times \mathbb{N} \to \mathbb{C}$ be a function. We note that

$$\operatorname*{Lim}_{x\to\infty} \operatorname*{Lim\,sup}_{y\to\infty} f(x,y) = 0$$

if and only if, for each $\varepsilon > 0$, there exists $m_0 \in \mathbb{N}$ such that, for each $m \geq m_0$, there exists $n(\varepsilon, m) \in \mathbb{N}$ such that

$$|f(m,n)| < \varepsilon \quad (n \geq n(\varepsilon, m))\,.$$

Let $f : X \times Y \to \mathbb{C}$ be bounded and continuous. Then there is a reformulation of the definition of 'f 0-clusters strongly' which is convenient. Let f, X, and Y be as in the definition. For each $\varepsilon > 0$ and $x \in X$, define

$$Y_{\varepsilon,x} = \{y \in Y : |f(x,y)| \geq \varepsilon\} ,$$

so that $Y_{\varepsilon,x}$ is a closed subset of Y. Next, for each $\varepsilon > 0$, define

$$X_\varepsilon = \{x \in X : Y_{\varepsilon,x} \quad \text{is not compact}\} .$$

Then it is easily checked that $\operatorname{Lim}_{x\to\infty} \operatorname{Lim}_{y\to\infty} f(x,y) = 0$ if and only if X_ε is compact for each $\varepsilon > 0$.

Suppose that f 0-clusters locally uniformly on $X \times Y$. Then certainly f 0-clusters strongly on $X \times Y$. The following trivial remark will be useful.

PROPOSITION 3.8. *Let X and Y be non-empty sets, and let*

$$f : X \times Y \to \mathbb{C}$$

be a bounded function. Suppose that f 0-clusters strongly on $X \times Y$. Then f 0-clusters on $X \times Y$. □

We note that the converse to the above proposition fails. To see this, take $X = Y = \mathbb{N}$, and let $\{S_n : n \in \mathbb{N}\}$ be a partition of $\mathbb{N}$, with each S_n infinite. Define $f : \mathbb{N} \times \mathbb{N} \to \mathbb{C}$ by setting $f(m,n) = 1$ whenever $m \in S_n$ and $f(m,n) = 0$ otherwise. Then, for each $m \in \mathbb{N}$, we have $\limsup_n f(m,n) = 1$, and so f does not 0-cluster strongly on $\mathbb{N} \times \mathbb{N}$. Now let (m_j) and (n_k) be sequences in $\mathbb{N}$, each consisting of distinct points. For each $k \in \mathbb{N}$, we have $f(m_j, n_k) = 1$ for at most one value of $j \in \mathbb{N}$, and so $\lim_j f(m_j, n_k) = 0$. Suppose that

$$\alpha := \lim_j \lim_k f(m_j, n_k)$$

exists. Either there exist $j_0, k_0 \in \mathbb{N}$ such that $y_k \in S_{j_0}$ $(k \geq k_0)$, in which case $f(m_j, n_k) = 0$ $(j > j_0,\ k \geq k_0)$ and $\alpha = 0$, or, for each $j \in \mathbb{N}$, only finitely many points n_k belong to S_j, in which case $\lim_k f(m_j, n_k) = 0$ $(j \in \mathbb{N})$, and again $\alpha = 0$. Hence f 0-clusters on $\mathbb{N} \times \mathbb{N}$.

We again give a small generalization of the above notations. Let $k \in \mathbb{N}$ with $k \geq 2$, let $X_1, \dots, X_k$ be non-empty sets, and let

$$f : X_1 \times \cdots \times X_k \to \mathbb{C}$$

be a function. Then f *0-clusters strongly on* $X_1 \times \cdots \times X_k$ if all k-fold limits of the form

$$\operatorname*{Lim}_{x_{\sigma(1)}} \cdots \operatorname*{Lim}_{x_{\sigma(k)}} f(x_{n_1}, \dots, x_{n_k})$$

are 0; here σ is one of the $k!$ permutations of $\mathbb{N}_k$.

DEFINITION 3.9. *Let A be a Banach algebra. A linear functional $\lambda \in A'$ is* almost periodic (*respectively,* weakly almost periodic) *if the map*

$$a \mapsto a \cdot \lambda, \quad A \to A',$$

is compact (respectively, weakly compact).

Specifically, an element λ of A' is weakly almost periodic if and only if the $\sigma(A', A'')$-closure S in A' of the set $A_{[1]} \cdot \lambda$ is $\sigma(A', A'')$-compact. Since S is Hausdorff in the $\sigma(A', A)$-topology, the weak and weak-$*$ topologies of A' agree on $A_{[1]} \cdot \lambda$. Also note that, by Mazur's theorem, S is the $\|\cdot\|$-closure of $A_{[1]} \cdot \lambda$ in this case.

DEFINITION 3.10. *The spaces of almost periodic and weakly almost periodic functionals on the Banach algebra A are denoted by $AP(A)$ and $WAP(A)$, respectively.*

Both $AP(A)$ and $WAP(A)$ are $\|\cdot\|$-closed A-submodules of A'; clearly, $AP(A) \subset WAP(A)$. Let $\varphi \in \Phi_A$. By (2.4), $A \cdot \varphi = \mathbb{C}\varphi$, and so

$$\Phi_A \subset AP(A)\,. \tag{3.1}$$

Let $\lambda \in A'$. It follows from Proposition 3.4, as pointed out in [Py1], that λ is weakly almost periodic if and only if

$$\lim_m \lim_n \langle a_m b_n,\, \lambda\rangle = \lim_n \lim_m \langle a_m b_n,\, \lambda\rangle \tag{3.2}$$

whenever (a_m) and (b_n) are sequences in $A_{[1]}$ and both repeated limits exist, and hence λ is weakly almost periodic if and only if $\lambda \circ m_A$ clusters on $A_{[1]} \times A_{[1]}$. In particular, $\lambda \in WAP(A)$ if and only if the map $a \mapsto \lambda \cdot a$, $A \to A'$, is weakly compact. We easily obtain the following characterization of $WAP(A)$, first given in [Py1].

PROPOSITION 3.11. *Let A be a Banach algebra, and let $\lambda \in A'$. Then we have $\lambda \in WAP(A)$ if and only if*

$$\langle \Phi \,\Box\, \Psi,\, \lambda\rangle = \langle \Phi \diamond \Psi,\, \lambda\rangle \quad (\Phi, \Psi \in A'')\,.$$

□

The following result is essentially [L4, Proposition 3.3].

PROPOSITION 3.12. *Let A be a Banach algebra with a bounded approximate identity. Then $WAP(A)$ and $AP(A)$ are neo-unital Banach A-bimodules, and*

$$AP(A) \subset WAP(A) \subset (A' \cdot A) \cap (A \cdot A').$$

PROOF. Take E to be $WAP(A)$ or $AP(A)$, so that E is a Banach A-bimodule.

Let (e_α) be a bounded approximate identity in A, say the bound is $m = \sup_\alpha \|e_\alpha\|$.

Let $\lambda \in E$, and take S to be the $\sigma(A', A'')$-closure of $A_{[m]} \cdot \lambda$, so that S is compact in the $\sigma = \sigma(A', A'')$-topology. The net $(e_\alpha \cdot \lambda)$ is contained in S, and so we may suppose that $(e_\alpha \cdot \lambda)$ converges in (S, σ), say $e_\alpha \cdot \lambda \xrightarrow{\sigma} \mu \in S$. For each $a \in A$, we have

$$\langle a,\, \mu\rangle = \lim_\alpha \langle a,\, e_\alpha \cdot \lambda\rangle = \lim_\alpha \langle ae_\alpha,\, \lambda\rangle = \langle a,\, \lambda\rangle$$

because (e_α) is a right approximate identity, and so $\mu = \lambda$. It follows that λ belongs to the σ-closure of S, and hence to $\overline{AE}$. Similarly, $\lambda \in \overline{EA}$. This show that E is an essential A-bimodule.

By Cohen's factorization Theorem 2.3(ii), the A-bimodule E is a neo-unital. The result follows. □

We shall see in Example 4.9 and in Chapter 5 various results that show the limits of the above theorem. For a further discussion of the spaces $AP(A)$ and $WAP(A)$, see [DuU].

PROPOSITION 3.13. *Let A be a Banach algebra, and let $\lambda \in A'$. Suppose that there are a reflexive space E and bounded linear operators $U : A \to E'$ and $V : A \to E$ such that*

$$\langle ab,\, \lambda\rangle = \langle Vb,\, Ua\rangle \quad (a, b \in A)\,.$$

Then λ is weakly almost periodic.

PROOF. Let (a_m) and (b_n) be two sequences in $A_{[1]}$ such that

$$\alpha := \lim_m \lim_n \langle a_m b_n,\, \lambda\rangle \quad \text{and} \quad \beta := \lim_m \lim_n \langle a_m b_n,\, \lambda\rangle$$

both exist. Then $\lim_m \lim_n \langle Vb_n,\, Ua_m\rangle$ and $\lim_m \lim_n \langle Vb_n,\, Ua_m\rangle$ both exist, and are equal to α and β, respectively. Since E is reflexive, the closed unit ball $E_{[1]}$ is weakly compact, and so $\alpha = \beta$ by Proposition 3.4. It follows that the equality (3.2) holds, and so $\lambda \in WAP(A)$. □

The unit ball $S := A_{[1]}$ is a semigroup, and each $\lambda \in A'$ defines an element of $\ell^\infty(S)$. Suppose that λ is almost periodic. Then $S \cdot \lambda$ is relatively compact in $(\ell^\infty(S), |\cdot|_S)$, and so $\lambda \cdot S$ is also relatively compact in $\{\ell^\infty(S), |\cdot|_S)$; see [BJM, p. 130]. In particular, we see that $\lambda \in AP(A)$ if and only if the map $a \mapsto \lambda \cdot a,\ A \to A'$, is compact.

The following characterization of Arens regularity is taken from [D, Theorem 2.6.17], [DuH], and [Pa2, 1.4.11], where more general versions are given; original sources include [Py1] and [Y1, Theorem 10].

THEOREM 3.14. *Let A be a Banach algebra. Then the following conditions on A are equivalent:*

(a) *A is Arens regular;*

(b) *for each $\Phi \in A''$, the map $L_\Phi : \Psi \mapsto \Phi \,\Box\, \Psi$ is continuous on the space $(A'', \sigma(A'', A'))$;*

(c) *for each $\lambda \in A'$, the function $\lambda \circ m_A$ clusters on $A_{[1]} \times A_{[1]}$;*

(d) *$WAP(A) = A'$.* □

We note the following standard corollary.

COROLLARY 3.15. *Let A be an Arens regular Banach algebra. Then closed subalgebras of A and quotients of A by a closed ideal are all also Arens regular.* □

COROLLARY 3.16. *Let A be a Banach algebra. Then $(A'', \Box\,)$ is a dual Banach algebra if and only if A is Arens regular.* □

CHAPTER 4

Examples

In this chapter, we shall describe a considerable number of specific examples; we shall give information about the products $\Box$ and $\Diamond$ on A'' for a Banach algebra A, and we shall also give a little information on continuous derivations from A. Most examples summarize and extend work that already exists in the literature.

There are various known results on the topological centres of related Banach algebras.

First, let A and B be Banach algebras, and let $C = A \oplus B$ as a Banach space, taking

$$\|(a, b)\| = \|a\| + \|b\| \quad (a \in A,\, b \in B),$$

for example. Then C is a Banach algebra for the product

$$(a_1, b_1)(a_2, b_2) = (a_1 a_2, b_1 b_2) \quad (a_1, a_2 \in A,\, b_1, b_2 \in B).$$

We have $C'' = A'' \oplus B''$. It is straightforward to check that

$$\begin{aligned}(\Phi_1, \Psi_1) \Box (\Phi_2, \Psi_2) &= (\Phi_1 \Box \Phi_2, \Psi_1 \Box \Psi_2)\,,\\ (\Phi_1, \Psi_1) \Diamond (\Phi_2, \Psi_2) &= (\Phi_1 \Diamond \Phi_2, \Psi_1 \Diamond \Psi_2)\,,\end{aligned}$$

for $\Phi_1, \Phi_2 \in A''$ and $\Psi_1, \Psi_2 \in B''$, and so

$$\mathfrak{Z}_t^{(j)}(C'') = \mathfrak{Z}_t^{(j)}(A'') \oplus \mathfrak{Z}_t^{(j)}(B'') \quad (j = 1, 2)\,. \tag{4.1}$$

Now let (A_n) be a sequence of Banach algebras, and set $\mathfrak{A} = c_0(A_n)$, so that $\mathfrak{A}$ is a Banach algebra for the coordinatewise operations (see [D, Example 2.1.18(iii)]). Then we identify $\mathfrak{A}'$ with $\ell^1(A_n')$ and $\mathfrak{A}''$ with $\ell^\infty(A_n'')$. It is easy to check that

$$(\Phi_n) \Box (\Psi_n) = (\Phi_n \Box \Psi_n) \quad \text{and} \quad (\Phi_n) \Diamond (\Psi_n) = (\Phi_n \Diamond \Psi_n)$$

for $(\Phi_n), (\Psi_n) \in \mathfrak{A}''$), and so

$$\mathfrak{Z}_t^{(j)}(\mathfrak{A}'') = \ell^\infty\left(\mathfrak{Z}_t^{(j)}(A_n'')\right) \quad (j = 1, 2)\,.$$

In particular, $\mathfrak{A}$ is Arens regular whenever each A_n is Arens regular.

Here is a related result, due to Ülger [U2]. Let A be a Banach algebra, and let Ω be a compact space. Then $\mathfrak{A} := C(\Omega, A)$ is also a

Banach algebra, as in [D, Example 2.1.18(iv)]. Suppose that A is Arens regular. Then $\mathfrak{A}$ is also Arens regular.

Again let (A_n) be a sequence of Banach algebras, and now consider the space $\mathfrak{A} = \ell^\infty(A_n)$, so that $\mathfrak{A}$ is a Banach algebra for the coordinatewise operations. Suppose that each algebra A_n is Arens regular. In general, it is not true that $\mathfrak{A}$ is also Arens regular; examples of a sequence (A_n) of finite-dimensional algebras such that $\ell^\infty(A_n)$ is not Arens regular are given in [Py2] and [PyU]. In Example 9.2, we shall give another example using the Beurling algebras that we are considering.

Finally in this area, we again let (A_n) be a sequence of Banach algebras, and now consider the space $\mathfrak{A} = \ell^1(A_n)$, so that $\mathfrak{A}$ is a Banach algebra for the coordinatewise operations. Then it is shown by Arikan in [Ark] that $\mathfrak{A}$ is Arens regular if and only if each Banach algebra A_n is Arens regular.

Let A and B be Banach algebras, and let $A \widehat{\otimes} B$ be their projective tensor product, as in [D, Theorem 2.1.22]. The question of the Arens regularity of $A \widehat{\otimes} B$ is studied in [U1]. Unfortunately, however, some of the results of [U1] are incorrect; we are indebted to Colin Graham for this information. Indeed, assertions 4.12 and 4.20–4.23 of [U1] are all false. To see this, take Ω to be a compact, totally disconnected set. Then $C(\Omega) \widehat{\otimes} C(\Omega)$ contains an isomorphic copy of the group algebra $L^1(D)$, where $D = \mathbb{Z}_2^{\mathbb{N}}$ is the Cantor group (see below and [GrmMc] for notation), as a closed subalgebra. Since $L^1(D)$ is not Arens regular, $C(\Omega) \widehat{\otimes} C(\Omega)$ cannot be Arens regular. It is proved in [U1, Corollary 4.17] that the Banach algebra $c_0 \widehat{\otimes} A$ is Arens regular for each C^*-algebra A.

Let Ω be an arbitrary infinite, compact space, and set

$$A = C(\Omega) \widehat{\otimes} C(\Omega)\,.$$

Then it has been proved by Colin Graham that the centre $\mathfrak{Z}(A'')$ contains a closed subalgebra isometrically isomorphic to $C(\Omega)'' \widehat{\otimes} C(\Omega)''$, and so A is not strongly Arens irregular.

We now give some more specific examples.

EXAMPLE 4.1. Let $A = \ell^1 = \ell^1(\mathbb{N})$, with pointwise product. Then A is a commutative, semisimple Banach algebra which is a dual Banach algebra with predual c_0. We have $((\ell^1)'', \Box\,) = \ell^1 \ltimes c_0^\circ$.

Take $a \in A$ and $\lambda \in A' = \ell^\infty$. Then, by calculation, $\lambda \cdot a,\ a \cdot \lambda \in c_0$. Now take $\Phi \in (\ell^1)''$, $a \in A$, and $\Psi \in c_0^\circ$. Then $\langle a \cdot \Psi, \lambda\rangle = \langle \Psi, \lambda \cdot a\rangle = 0$ for each $\lambda \in \ell^\infty$, and so $a \cdot \Psi = 0$. Thus $\Phi \,\Box\, \Psi = 0$. Similarly $\Psi \cdot a = 0$.

It follows that the product $\Box$ in $\ell^1 \ltimes c_0^\circ$ is specified by

$$(a,\,\Phi)\,\Box\,(b,\,\Psi) = (ab,\,0) \quad (a, b \in A,\, \Phi, \Psi \in c_0^\circ)\,.$$

This shows that $(A'',\,\Box\,)$ is a commutative Banach algebra and that A is Arens regular; by Proposition 2.16, we have $\mathrm{rad}(A'',\,\Box\,) = c_0^\circ$.

In fact, we can identify ℓ^∞ with $C(\beta\mathbb{N})$ and A'' with $M(\beta\mathbb{N})$, as in Chapter 1, so that $c_0^\circ = M(\beta\mathbb{N} \setminus \mathbb{N})$. In particular, A is a closed ideal in $(A'',\,\Box\,)$.

This examples originates with Civin and Yood [CiY]; for details, see [D, Example 2.6.22(iii)]. $\Box$

EXAMPLE 4.2. Let A be a C^*-algebra. Then the algebra A has a $*$-representation as a closed and $*$-closed subalgebra of $\mathcal{B}(H)$, where H is a Hilbert space. Suppose that every state on $A^\#$ has the form

$$T \mapsto [Tx, x], \quad A^\# \to \mathbb{C}\,,$$

for some $x \in H$. Then the $*$-representation of A is said to be *universal*. Every C^*-algebra has a universal $*$-representation. In this case, A'' is identified with the von Neumann algebra which is the second commutant of A in $\mathcal{B}(H)$; this latter algebra is also equal to the closure of A in both the weak operator and strong operator topologies, and is called the *enveloping von Neumann algebra* of A. See [D, Chapter 3.2], and [Ta, III.2] for details.

It is standard (e.g., [D, Theorem 3.2.36]) that both of the products $\Box$ and $\Diamond$ on A'' coincide with the given product in $\mathcal{B}(H)$ for elements of A'', and that the linear involution that we have specified on A'' coincides with the restriction of the involution on $\mathcal{B}(H)$. In particular, every C^*-algebra A is Arens regular, and $(A'', \Box)$ is also a C^*-algebra.

The first proofs of the fact that each C^*-algebra is Arens regular were given by Sherman [Sh] and by Takeda [Td]; see also [CiY].

A different method of showing that a C^*-algebra is Arens regular and that the second dual is a C^*-algebra uses the Vidav–Palmer theorem; see [BoDu, Theorem 38.19].

It is even true that a Banach algebra A which is a C^*-algebra with respect to a different product is still Arens regular: this follows from a theorem of Akemann [Ak] that every continuous linear map from a C^*-algebra A into A' is weakly compact, and so $WAP(A) = A'$. Thus A is Arens regular by Theorem 3.14.

Let H be a Hilbert space. Then it follows from Corollary 3.15 that every closed subalgebra of $\mathcal{B}(H)$ is also Arens regular. See also [ER] for this remark.

Let $A = C_0(\Omega)$ be the commutative C^*-algebra of all continuous functions that vanish at infinity on a non-empty locally compact space

Ω. Then A'' is the algebra $C(\widetilde{\Omega})$ for some extremely disconnected, compact space $\widetilde{\Omega}$ [D, Theorem 4.2.29]. In particular,

$$c_0'' = \ell^\infty = C(\beta\mathbb{N})\,.$$

There is a massive theory which explains when a C^*-algebra is amenable; for a summary with references, see [D, Chapter 5.7], and, for a clear, stream-lined account, see [Ru2, Chapter 6]. A C^*-algebra is always weakly amenable (see [D, Theorem 5.6.77], where the proof is taken from [HaLaus]) and 2-weakly amenable [DGhGr]. □

EXAMPLE 4.3. Let G be a locally compact group, and let $(L^1(G),\,\star\,)$ be the group algebra of G (see below). Then $L^1(G)$ is a Banach algebra which is Arens regular if and only if G is finite. This was proved for abelian groups G by Civin and Yood [CiY] and in the general case by Young [Y2]. Indeed $L^1(G)$ is always strongly Arens irregular. This was first proved in the case where G is abelian by M. Grosser and Losert [GLos] (see also Parsons [Par]). It was then proved in the case where G is compact in [IPyU], and in the general case in [LLos1]; see also [LU, Corollary 5.5 and p. 1210], where we note that Lemma 5.3 of [LU] is not quite precise, and that Theorem 5.4 may not be true. A new and shorter proof of a slightly stronger result has been given recently by Neufang [N1]. We shall prove a generalization of the result in Theorem 12.3.

In [Si], the group algebra $(L^1(G),\,\star\,)$ is considered with different topologies with respect to which it is a locally convex algebra, and related results are obtained.

It follows from Corollary 3.15 that, in the case where G is infinite, the group algebra $L^1(G)$ is never a closed subalgebra or the quotient of a C^*-algebra.

Let $A = \ell^1(\mathbb{Z})$, identified with $A(\mathbb{T}) \subset C(\mathbb{T})$ in the standard way. Let E be a Helson subset of $\mathbb{T}$, so that B, defined to be

$$\{f \mid E : f \in A(\mathbb{T})\}\,,$$

is equal to $C(E)$. Then A is strongly Arens irregular, but its quotient B is Arens regular, and so the inclusion in (2.24) is strict in this case.

It is a famous theorem of Johnson [J1] that the Banach algebra $L^1(G)$ is amenable if and only if the locally compact group G is amenable (see Definition 7.36, below). It is a further theorem of Johnson [J2] that the Banach algebra $L^1(G)$ is always weakly amenable; for a shorter proof, due to Despic and Ghahramani [DesGh], see [D, Theorem 5.6.48]. It is not known whether or not $L^1(G)$ is always 2-weakly amenable; this is certainly true if G is an amenable group, and it is true when G is the (non-amenable, discrete) group $\mathbb{F}_2$, the free group on two generators [J3].

An algebra closely related to the group algebra $L^1(G)$ is the *measure algebra* $M(G)$ of G (see Chapter 7, below).

The question whether or not $M(G)$ is strongly Arens irregular was raised in [GhL2]. This was partially solved by Neufang in [N4]: this is the case whenever G is a locally compact, non-compact group with $\kappa(G)$ (see Definition 6.39) a non-measurable cardinal (see Chapter 11).

It has recently been determined when $M(G)$ is amenable and when it is weakly amenable [DGhH]: $M(G)$ is amenable if and only if the locally compact group G is discrete and amenable, and $M(G)$ is weakly amenable if and only if G is discrete. □

The above two examples exhibit the 'extreme' cases in the inclusions of (2.22). It is easy to find intermediate cases; more natural examples of the intermediate situation will be given in Examples 6.2 and 9.7, below.

EXAMPLE 4.4. Let A and B be Banach algebras, and set $C = A \oplus B$, as above, so that $\mathfrak{Z}_t^{(j)}(C'') = \mathfrak{Z}_t^{(j)}(A'') \oplus \mathfrak{Z}_t^{(j)}(B'')$ $(j = 1, 2)$. By taking A to be Arens regular and B to be strongly Arens irregular, we obtain algebras C which are neither Arens regular nor strongly Arens irregular.

Let A be a Banach $*$-algebra. We have noted that the involution extends to a linear involution on A'' that maps $\mathfrak{Z}_t^{(1)}(A'')$ onto $\mathfrak{Z}_t^{(2)}(A'')$. The following is the basis of an example that will show in due course that we can have $\mathfrak{Z}_t^{(1)}(A'') \neq \mathfrak{Z}_t^{(2)}(A'')$; see Examples 6.3 and 10.13.

Let A be a Banach algebra, and suppose that there is a linear involution $a \mapsto \overline{a}$ on A such that

$$\overline{ab} = \overline{a}\,\overline{b} \quad (a, b \in A)\,.$$

It is easily checked that this linear involution extends to a linear involution $\Phi \mapsto \overline{\Phi}$ on A'' such that $\overline{\Phi \,\square\, \Psi} = \overline{\Phi} \,\square\, \overline{\Psi}$ $(\Phi, \Psi \in A'')$. Thus $\overline{\Phi} \in \operatorname{rad}(A'', \,\square\,)$ if and only if $\Phi \in \operatorname{rad}(A'', \,\square\,)$.

Set $C = A \oplus A^{\mathrm{op}}$, and, for $a, b \in A$, define

$$(a, b)^* = (\overline{b}, \overline{a})\,.$$

We *claim* that $*$ is an involution on C. Certainly $*$ is a linear involution. Further, for $a_1, a_2, b_1, b_2 \in A$, we have

$$((a_1,\, b_1)(a_2,\, b_2))^* = (a_1 a_2,\, b_2 b_1)^* = (\overline{b}_2\, \overline{b}_1, \overline{a}_1\, \overline{a}_2)$$

and

$$(a_2,\, b_2)^*(a_1,\, b_1)^* = (\overline{b}_2,\, \overline{a}_2)(\overline{b}_1, \overline{a}_1) = (\overline{b}_2\, \overline{b}_1, \overline{a}_1\, \overline{a}_2)\,,$$

and so $*$ is indeed an involution. The algebra $(C, *)$ is a Banach $*$-algebra. As before, the involution extends to a linear involution on C''; indeed, we have $(\Phi, \Psi)^* = (\overline{\Psi}, \overline{\Phi})$ $(\Phi, \Psi \in A'')$.

It is clear that

$$\mathfrak{Z}_t^{(1)}(C'') = \mathfrak{Z}_t^{(1)}(A'') \oplus \mathfrak{Z}_t^{(2)}(A'') \quad \text{and} \quad \mathfrak{Z}_t^{(2)}(C'') = \mathfrak{Z}_t^{(2)}(A'') \oplus \mathfrak{Z}_t^{(1)}(A'')\,.$$

Thus, in the case where $\mathfrak{Z}_t^{(1)}(A'') \neq \mathfrak{Z}_t^{(2)}(A'')$, we shall obtain a Banach $*$-algebra C such that $\mathfrak{Z}_t^{(1)}(C'') \neq \mathfrak{Z}_t^{(2)}(C'')$.

Temporarily, we define $B = A^{\mathrm{op}}$, $R = \mathrm{rad}\,(A'', \Box\,) = \mathrm{rad}\,(B'', \Diamond\,)$, and $S = \mathrm{rad}\,(B'', \Box\,) = \mathrm{rad}\,(A'', \Diamond\,)$. Then we see that

$$\mathrm{rad}\,(C'', \Box\,) = R \oplus S, \quad \mathrm{rad}\,(C'', \Diamond)\,) = S \oplus R\,.$$

It follows that $\mathrm{rad}\,(C'', \Box\,) \neq \mathrm{rad}\,(C'', \Diamond)$ whenever $S \neq R$; we shall give examples of this situation in Example 6.3. It is reassuring that our calculations show that $(\Phi, \Psi) \in \mathrm{rad}\,(C'', \Box\,)$ if and only if we have $(\Phi, \Psi)^* \in \mathrm{rad}\,(C'', \Diamond\,)$, as predicted earlier. □

The following example is based on one given in [GhMMe].

EXAMPLE 4.5. Let A be a Banach algebra, and let E be a Banach A-bimodule. Then the Banach space

$$\mathfrak{A} = A \oplus E$$

is a Banach algebra for the product specified by

$$(a, x)(b, y) = (ab,\ a \cdot y + x \cdot b) \quad (a, b \in A,\ x, y \in E)\,.$$

For details of this standard construction, see [D, p. 39]. The topological centre $\mathfrak{Z}_t^{(1)}(\mathfrak{A}'')$ is identified in [GhMMe]. Indeed, it consists of the elements of the form $(\Phi, \Lambda) \in \mathfrak{A}'' = A'' \oplus E''$ such that:

(i) $\Phi \in \mathfrak{Z}_t^{(1)}(A'')$;

(ii) the map $\mathrm{M} \mapsto \Phi \cdot \mathrm{M}$ is continuous on $(E'', \sigma(E'', E'))$;

(iii) the map $\Psi \mapsto \Lambda \cdot \Psi$ is continuous from the space $(A'', \sigma(A'', A'))$ to $(E'', \sigma(E'', E'))$.

Now suppose that A is left strongly Arens irregular, so that we have $\mathfrak{Z}_t^{(1)}(A'') = A$. Then $\mathfrak{Z}_t^{(1)}(\mathfrak{A}'')$ consists of the elements $(a, \Lambda) \in A \oplus E''$ such that the map $\Psi \mapsto \Lambda \cdot \Psi$ is continuous from $(A'', \sigma(A'', A'))$ to $(E', \sigma(E'', E'))$. In particular, in the case where $EA = \{0\}$, so that $\Lambda \cdot \Psi = 0$ for each $\Lambda \in E''$ and $\Psi \in A''$, we have $\mathfrak{Z}_t^{(1)}(\mathfrak{A}'') = A \oplus E''$. However, in the case where $E = A$ and the left module operation on E is the product in A, the map $\Psi \mapsto \Lambda \cdot \Psi = \Lambda \,\Box\, \Psi$ is continuous if and only if $\Phi \in A$, and so $\mathfrak{Z}_t^{(1)}(\mathfrak{A}'') = A \oplus E'' = \mathfrak{A}$.

We now suppose that A is strongly Arens irregular, and take $E = A$ as a Banach space, setting

$$a \cdot x = ax, \ x \cdot a = 0 \quad (a \in A, x \in E).$$

Then we have seen that $\mathfrak{Z}_t^{(1)}(\mathfrak{A}'') = \mathfrak{A}$, and so $\mathfrak{A}$ is left strongly Arens irregular.

The product in the opposite algebra A^{op} is given by

$$(a, x) \cdot (b, y) = (ba, bx) \quad (a, b, x, y \in A),$$

and so $\mathfrak{Z}_t^{(1)}(\mathfrak{A}'') = \mathfrak{Z}_t^{(1)}((A^{\mathrm{op}})'') = A \oplus A''$. In this case $\mathfrak{A}$ is not strongly Arens regular whenever A is not reflexive as a Banach space.

Thus we can take $A = (L^1(G), \star)$ for an infinite locally compact group G, and set $\mathfrak{A} = A \oplus A$, with the above product, to obtain a Banach algebra that is left strongly Arens irregular, but that is not right strongly Arens irregular. □

An earlier example of a Banach algebra $\mathfrak{A}$ which is left strongly Arens irregular, but that is not right strongly Arens irregular, was given by Neufang [N3]. The algebra $\mathfrak{A}$ is, as a Banach space, the set $\mathcal{N}(L^p(G))$ of nuclear operators on the Banach space $L^p(G)$ for a locally compact group G and an index p with $1 < p < \infty$. However Neufang introduces a product on $\mathfrak{A}$ which is a form of convolution and which is different from the usual composition of operators. The algebra $\mathfrak{A}$ is a Banach algebra for this new product, and $\mathfrak{A}$ has a bounded right approximate identity of norm 1. There is a canonical quotient map from $\mathfrak{A}$ onto the group algebra $(L^1(G), \star)$, say this map has kernel the closed ideal I. We have $\mathfrak{A}I = 0$. It is shown by Neufang that, in the case where G is a non-compact, second countable locally compact group, the algebra $\mathfrak{A}$ is left strongly Arens irregular, but not right strongly Arens irregular. The work of Neufang develops interesting properties of the algebra $\mathfrak{A}$ as a non-commutative convolution algebra; a detailed study of the homological properties of this algebra has recently been given by Pirkovskii in [Pir].

EXAMPLE 4.6. Let B be a unital C^*-algebra. Take U to be the unitary group of B, and set $A = \ell^1(U)$, the group algebra of U. The map

$$q : \sum \alpha_u \delta_u \mapsto \sum \alpha_u u, \quad A \to B,$$

is a continuous linear map. Since every element of B is a linear combination of 4 unitary elements, q is a surjection. For $i = 1, 2$, we have $\mathfrak{Z}_t^{(i)}(A'') = A$ and $\mathfrak{Z}_t^{(i)}(B'') = B''$, and so the inclusion in (2.24) is strict in this case. □

EXAMPLE 4.7. Let A be the *Volterra algebra* $(L^1(\mathbb{I}), \star)$ or a weighted convolution algebra $(L^1(\mathbb{R}^+, \omega), \star)$, where ω is a regulated weight on $\mathbb{R}^+$. (For definitions, see [D, Chapter 4.7].) Then A is strongly Arens irregular. This was first proved by Ghahramani and McClure in [GhM]; see also [LU, Corollary 3.5].

One can see easily that, with A equal to the Volterra algebra, which has a sequential bounded approximate identity, we have

$$A' \cdot A = \overline{A'A} = C_0[0,1)$$

and that A is a closed ideal in A''.

On the other hand, let A be the algebra $C_\star$; this is the Banach space $C(\mathbb{I})$ with convolution multiplication of functions, as in [D, Definition 4.7.39]. Then A is a radical Banach algebra, and, by a remark in Example 4.2, A is Arens regular. □

EXAMPLE 4.8. Let A be a Banach algebra which is reflexive as a Banach space. (For example, let A be the Banach space ℓ^p, where $1 < p < \infty$, taken with coordinatewise multiplication: for details of this example, see [D, Example 4.1.42].) Then A is both Arens regular and strongly Arens irregular. □

EXAMPLE 4.9. (i) Let A be a non-zero Banach algebra with zero product, so that $A^2 = 0$. Then both $\Box$ and $\Diamond$ are the zero product on A'', and so A is Arens regular. Clearly $AP(A) = WAP(A) = A'$ and $A' \cdot A = A \cdot A' = 0$, and so neither $AP(A)$ nor $WAP(A)$ is essential (*cf.* Proposition 3.12).

(ii) Let $A = \mathbb{C}^2$, with the product given by

$$(z_1, z_2)(w_1, w_2) = (z_1 w_1, z_1 w_2) \quad (z_1, z_2, w_1, w_2 \in \mathbb{C})\,.$$

Then A is a Banach algebra. Set $p = (1, 0)$ and $q = (0, 1)$. Then we have $pa = a \ (a \in A)$, and so p is a left identity for A. Clearly

$$AP(A) = WAP(A) = A' = \mathbb{C}^2\,.$$

For each $\lambda \in A'$, we have $\lambda \cdot p = \lambda$, and so $A = A' \cdot A$.

Let $\mu = (0, 1) \in A'$. Then $\langle q,\, \mu\rangle = 1$, but

$$\langle q,\, a \cdot \lambda\rangle = \langle qa, \lambda\rangle = 0 \quad (a \in A,\, \lambda \in A')\,,$$

and so $\mu \notin A \cdot A'$. Hence $A \cdot A' \subsetneq A' = WAP(A)$. This shows that we cannot replace 'bounded approximate identity' by 'bounded left approximate identity' in Proposition 3.12. □

EXAMPLE 4.10. Let J be the *James space*, so that J is a certain Banach space of sequences on $\mathbb{N}$. In fact, J is a non-unital Banach algebra with respect to coordinatewise multiplication, and J has a bounded approximate identity; the algebra $J^{\#}$ is identified with $J \oplus \mathbb{C}1$, where $1 = (1, 1, \dots)$. In this case, J'' is isometrically isomorphic to $J^{\#}$ (when J has a suitable norm), and the two products $\Box$ and $\Diamond$ coincide with the natural product on $J^{\#}$. Thus J is Arens regular.

For further details of this example, see [D, Example 4.1.45]. $\Box$

EXAMPLE 4.11. Let G be a locally compact group, and take p such that $1 < p < \infty$. Then $A_p(G)$ denotes the *Herz algebra*, described in [D, Theorem 4.5.30]; $A_p(G)$ is a Banach function algebra on G. In the case where $p = 2$, $A(G) = A_2(G)$ is the *Fourier algebra* of G.

The Arens regularity of these algebras was first considered in [LW], where it was shown that $A_2(G)$ is not Arens regular for each infinite, amenable group G. In fact, it is probable that the algebras $A_p(G)$ are Arens regular only if G is finite. Towards this, Forrest [Fo1], [Fo2] has shown that, if $A_p(G)$ is Arens regular, then G must be discrete, and that, if $A(G)$ is Arens regular, then G cannot contain an infinite, amenable subgroup; the latter remark is also proved in a diferent way in [U5, Corollary 3.7]. (However, there exist infinite, non-amenable groups for which all proper subgroups are finite; for this, see [Ol1] and [Ol2].) In the case where G is not discrete, various quotients of $A_p(G)$ are not Arens regular [Gra4, Corollary 8].

In fact, it is proved in [LLos2, Theorem 6.5] that, for discrete, amenable groups G, the Fourier algebra $A(G)$ is strongly Arens irregular.

Let E be a closed subset of G, and denote by $A_p(E)$ the restriction algebra, $A_p(G) \mid E$, so that $A_p(E)$ is a Banach function algebra on E. It is shown by Granirer [Gra4] and Graham [Grm1], [Grm3] that in various cases $A_p(E)$ is not Arens regular.

The Arens regularity of some related algebras is considered by Granirer in [Gra6]. $\Box$

EXAMPLE 4.12. Let A be an Arens regular Banach algebra. In general, the second dual $(A'', \Box)$ is not necessarily itself Arens regular. For example, Pym gives in [Py2] an example of an Arens regular Banach function algebra A such that $(A'', \Box)$ is not Arens regular. In [Grm2], Graham constructs various quotient algebras $A(E)$ of the Fourier algebras $A(G)$ such that $A(E)$ is Arens regular, but $A(E)''$ is not Arens regular (and such that $A(E)$ has various additional properties). See also [Grm4]. $\Box$

EXAMPLE 4.13. We have noted in Example 4.3 that the Banach algebra $(L^1(\mathbb{T}), \star)$ is strongly Arens irregular. Recall that

$$H^1(\mathbb{T}) = \{f \in L^1(\mathbb{T}) : \widehat{f}(-n) = 0 \ \ (n \in \mathbb{N})\},$$

where $\widehat{f}$ denotes the Fourier transform of f. Thus $H^1(\mathbb{T})$ is a closed ideal in $L^1(\mathbb{T})$. It has been proved by Colin Graham that $H^1(\mathbb{T})$ is Arens regular, as are $H^1(\mathbb{T}) \widehat{\otimes} H^1(\mathbb{T})$ and all the even duals of $H^1(\mathbb{T})$. Thus an infinite-dimensional closed ideal in a strongly Arens irregular Banach function algebra might itself be Arens regular. □

EXAMPLE 4.14. In [U5], Ülger studies Banach function algebras A which are weakly sequentially complete as Banach spaces and which are such that L_f is a weakly compact operator on A for each $f \in A$, so that A is a closed ideal in $(A'', \Box)$. For example, $A(G)$ has all these properties whenever G is a discrete group [L2, Theorem 3.7].

Let A be such an algebra. Then it is shown by Ülger that

$$\mathfrak{Z}(A'') = \{\Phi \in A'' : \Phi \Box A'', A'' \Box \Phi \subset A\}.$$

Suppose, further, that A has a bounded approximate identity. Then A is strongly Arens irregular. □

CHAPTER 5

Introverted Subspaces

Let A be a Banach algebra. We shall now consider certain subspaces of A' which give rise to quotient algebras of $(A'', \Box)$.

DEFINITION 5.1. *Let A be a Banach algebra, and let X be a $\|\cdot\|$-closed, A-submodule of A'. Then X is:*

(i) faithful *if $a = 0$ whenever $a \in A$ and $\langle a, \lambda \rangle = 0$ $(\lambda \in X)$;*

(ii) left-introverted *if $\Phi \cdot \lambda \in X$ $(\lambda \in X, \Phi \in A'')$.*

Our definition of 'faithful' is slightly weaker than the classical definition, which is that $a = 0$ whenever $a \cdot \lambda = 0$ for each $\lambda \in X$; it is equivalent to the classical definition for algebras A such that $a = 0$ whenever $Aa = \{0\}$.

PROPOSITION 5.2. *Let A be a Banach algebra, and let X be a $\|\cdot\|$-closed, A-submodule of A'. Then X is left-introverted if and only if $\Phi \cdot \lambda \in X$ whenever $\lambda \in X$ and $\Phi \in X'$.*

PROOF. This follows easily from the Hahn–Banach theorem. $\Box$

For example, the subspaces A' and $\overline{A'A}$ are both left-introverted subspaces of A'; the space A' is always faithful, and $\overline{A'A}$ is faithful whenever $Aa \neq \{0\}$ for each $a \in A^{\bullet}$. For further examples, see Theorem 7.19.

The notion of left-introversion was introduced in a special case by Day in [Day]; see also [Wo] and [LLo2]. We note that our definition of 'left-introverted' generalizes that of [BJM, Definition 2.4] and is different from that in [Pat, (2.6)].

The following characterization of left-introverted spaces is given in [LLo2, Lemma 1.2].

PROPOSITION 5.3. *Let A be a Banach algebra, and let X be a $\|\cdot\|$-closed, A-submodule of A'. Then X is left-introverted if and only if the $\sigma(A', A)$-closure in A' of $A_{[1]} \cdot \lambda$ is contained in X for each $\lambda \in X$.* $\Box$

As above, $X^{\circ} = \{\Phi \in A'' : \langle \Phi, \lambda \rangle = 0 \ (\lambda \in X)\}$ for a closed linear subspace X of A'. Clearly X° is a closed linear subspace of A''.

THEOREM 5.4. *Let A be a Banach algebra, and let X be a $\|\cdot\|$-closed A-submodule of A'.*

(i) *The space X° is a closed left ideal in $(A'', \Box\,)$.*

(ii) *Suppose, further, that X is left-introverted. Then X° is a closed ideal in $(A'', \Box\,)$.*

(iii) *Suppose, further, that $A' \cdot A \subset X$. Then X° is a left-annihilator ideal in $(A'', \Box\,)$ and $X^\circ \subset \operatorname{rad}(A'', \Box\,)$.*

(iv) *Suppose that $\overline{A'A} \neq A'$. Then $\operatorname{rad}(A'', \Box\,) \neq \{0\}$.*

PROOF. Let $\Phi \in X^\circ$ and $\Psi \in A''$.

(i) For each $a \in A$ and $\lambda \in X$, we have $\langle a,\, \Phi \cdot \lambda\rangle = \langle \Phi,\, \lambda \cdot a\rangle = 0$ because $\lambda \cdot a \in X$, and so $\Phi \cdot \lambda = 0$. Thus $\langle \Psi \Box \Phi,\, \lambda\rangle = 0$, and so $\Psi \Box \Phi \in X^\circ$. Hence X° is a left ideal in $(A'', \Box\,)$.

(ii) For each $\lambda \in X$, we have

$$\langle \Phi \Box \Psi,\, \lambda\rangle = \langle \Phi,\, \Psi \cdot \lambda\rangle = 0$$

because $\Psi \cdot \lambda \in X$, and so $\Phi \Box \Psi \in X^\circ$. Hence X° is a right ideal in $(A'', \Box\,)$.

(iii) Let $\lambda \in A'$. For each $a \in A$, we have $\langle a, \Phi \cdot \lambda\rangle = \langle \Phi, \lambda \cdot a\rangle = 0$ because $\lambda \cdot a \in X$ by hypothesis, and so $\Phi \cdot \lambda = 0$. Thus $\langle \Psi \Box \Phi, \lambda\rangle = 0$, and hence $\Psi \Box \Phi = 0$ in A''. This shows that X° is a left-annihilator ideal, and hence that $X^\circ \subset \operatorname{rad}(A'', \Box\,)$.

(iv) Set $E = \overline{A'A}$ (where we take the closure in $(A', \|\cdot\|)$). Then we have $E^\circ \neq \{0\}$, and so this is immediate from (iii). $\Box$

Let X be a left-introverted submodule of A'. As a Banach space, we have $X' = A''/X^\circ$, and so we can regard X' as a quotient Banach algebra of $(A'', \Box\,)$. The product in X' is again denoted by $\Box$, so that $\Phi \Box \Psi$ is defined in X' (for $\Phi, \Psi \in X'$) by the formula:

$$\langle \Phi \Box \Psi,\, \lambda\rangle = \langle \Phi,\, \Psi \cdot \lambda\rangle \quad (\lambda \in X)\,. \tag{5.1}$$

There is a natural map of A into X'; in the case where X is faithful, the map is an embedding, and we regard A as a subalgebra of $(X', \Box\,)$. Again, $a \cdot \Phi = a \Box \Phi$ and $\Phi \cdot a = \Phi \Box a$ for each $a \in A$ and $\Phi \in X'$. As usual, we define operators L_Φ and R_Φ in $\mathcal{B}(X')$ for $\Phi \in X'$ by the formulae:

$$L_\Phi(\Psi) = \Phi \Box \Psi, \quad R_\Phi(\Psi) = \Psi \Box \Phi \quad (\Psi \in X')\,.$$

The topology on X' is taken to be the weak-$*$ topology, $\sigma(X', X)$, unless we state otherwise. It is easy to see that each operator R_Φ is continuous on X'.

DEFINITION 5.5. *Let A be a Banach algebra, and let X be a left-introverted subspace of A'. Then the* topological centre *of X' is*

$$\mathfrak{Z}_t(X') = \{\Phi \in X' : L_\Phi \text{ is continuous on } (X', \sigma(X', X))\}.$$

Let $\Phi \in X'$. Then, by (5.1), $\Phi \in \mathfrak{Z}_t(X')$ if and only if the linear functional

$$\Psi \mapsto \langle \Phi, \Psi \cdot \lambda \rangle, \quad X' \to \mathbb{C},$$

is continuous on X' for each $\lambda \in X$. Clearly the set $\mathfrak{Z}_t(X')$ is a $\|\cdot\|$-closed subalgebra of $(X', \Box)$, and, when X is faithful,

$$A \subset \mathfrak{Z}_t(X') \subset X'. \tag{5.2}$$

The space $\mathfrak{Z}_t(X')$ coincides with $\mathfrak{Z}_t^{(1)}(A'')$, as previously defined in Definition 2.17, in the case where $X = A'$. If A is a commutative Banach algebra, then $\mathfrak{Z}_t(X') = \mathfrak{Z}(X')$, the centre of the algebra $(X', \Box)$. Suppose that A is Arens regular. Then $\mathfrak{Z}_t(X') = X'$ for each left-introverted subspace of A'.

The notion of the topological centre $\mathfrak{Z}_t(X')$ in the above sense was introduced in [IPyU]. (In the case where $X = \overline{A'A}$, the space $\mathfrak{Z}_t(X')$ was denoted by $\widetilde{Z}_1$ in [LU].)

Again let X be $\|\cdot\|$-closed, A-submodule of A'. Then we say that X is *right-introverted* if

$$\lambda \cdot \Phi \in X \quad (\lambda \in X,\ \Phi \in A'').$$

Let X be such a space. Then X° is a closed ideal in the Banach algebra $(A'', \Diamond)$, and we can identify X' with the quotient Banach algebra $(A'', \Diamond)/X'$; the product in X' is denoted by $\Diamond$, so that

$$\langle \Phi \Diamond \Psi, \lambda \rangle = \langle \Psi, \lambda \cdot \Phi \rangle \quad (\lambda \in X). \tag{5.3}$$

Suppose, further, that $A \cdot A' \subset X$. Then X° is a right-annihilator ideal, and $X^\circ \subset \operatorname{rad}(A'', \Diamond)$.

DEFINITION 5.6. *Let A be a Banach algebra, and let X be a $\|\cdot\|$-closed, A-submodule of A'. Then X is* introverted *if it is both left-introverted and right-introverted.*

In this case, the topological centre of X' is

$$\mathfrak{Z}_t(X') = \{\Phi \in X' : \Phi \Box \Psi = \Phi \Diamond \Psi \quad (\Psi \in X')\}.$$

Suppose that A is commutative. Then a $\|\cdot\|$-closed, A-submodule X of A' is introverted whenever it is left-introverted.

For example, consider the Banach algebra $A = (\ell^1, \cdot)$, so that the dual modules are $A' = C(\beta\mathbb{N})$ and $A'' = M(\beta\mathbb{N})$, as in Example 4.1. For a closed subset S of $\beta\mathbb{N}$, set

$$X = \{\lambda \in C(\beta\mathbb{N}) : \lambda \mid S = 0\},$$

a closed ideal in $C(\beta\mathbb{N})$. Then X is an introverted subspace of A' and $X' = M(\beta\mathbb{N} \setminus S)$. Clearly $\mathfrak{Z}_t(X') = X'$.

For further examples of introverted subspaces, see Chapter 11.

PROPOSITION 5.7. *Let A be a Banach algebra, and let X be a $\|\cdot\|$-closed, A-submodule of A' with $X \subset WAP(A)$. Then X is introverted.*

PROOF. Let $\lambda \in X$, and take S to be the $\sigma(A', A'')$-closure of $A_{[1]} \cdot \lambda$. Then, as we remarked, S is the $\|\cdot\|$-closure of $A_{[1]} \cdot \lambda$ because λ is weakly almost periodic, and so $S \subset X$. Also, S is the $\sigma(A', A)$-closure of $A_{[1]} \cdot \lambda$, and so, by Proposition 5.3, X is left-introverted.

Similarly, X is right introverted. □

COROLLARY 5.8. *Let A be a Banach algebra. Then $WAP(A)$ is an introverted subspace of A'.* □

We now make some remarks that are essentially contained in [LU, Chapter 3] and [BaLPy]. The results were originally due to M. Grosser [G1]; in particular, see Satz 4.14 and pages 181–182 of [G1].

Let A be a Banach algebra with a bounded approximate identity (e_α), and let Φ_0 be a corresponding mixed identity. Then L_{Φ_0} is a continuous projection on $(A'', \|\cdot\|)$; by an earlier remark, L_{Φ_0} is a homomorphism on $(A'', \Box)$, and so

$$A'' = \Phi_0 \Box A'' \ltimes (1 - \Phi_0) \Box A''$$

as a semidirect product and also as a direct sum of closed subspaces. The set

$$X := A' \cdot A = \{\lambda \cdot a : a \in A,\, \lambda \in A'\} = \overline{A'A}$$

is a left-introverted subspace of A'. Let $\Phi \in A''$. Then $\Phi \in X^\circ$ if and only if $\langle \Phi, \lambda \cdot a\rangle = 0$ $(a \in A,\, \lambda \in A')$, and so $\Phi \in X^\circ$ if and only if $\Phi \cdot \lambda = 0$ $(\lambda \in A')$. On the other hand, $\Phi \in (1 - \Phi_0) \Box A''$ if and only if $\Phi_0 \Box \Phi = 0$, and hence if and only if $\lim_\alpha \langle e_\alpha, \Phi \cdot \lambda\rangle = 0$ $(\lambda \in A')$; this occurs if and only if $\Phi \cdot \lambda = 0$ $(\lambda \in A')$. Thus we conclude that $X^\circ = (1 - \Phi_0) \Box A''$.

We state the above result as a proposition; a special case of the result is contained in [GhL1].

PROPOSITION 5.9. *Let A be a Banach algebra such that A'' has a mixed identity Φ_0. Then $X := A' \cdot A$ is a left-introverted subspace of A', $(A'', \Box)$ is the semidirect product*

$$A'' = X' \ltimes X^\circ ,$$

and the Banach space X' is linearly homeomorphic to $\Phi_0 \Box A''$. Further, $\mathfrak{Z}_t^{(1)}(A'') \cap X' \subset \mathfrak{Z}_t(X')$.

PROOF. Let $\Phi \in \mathfrak{Z}_t^{(1)}(A'') \cap X'$. Then the map $L_\Phi : \Psi \mapsto \Phi \Box \Psi$ is continuous on A'', and hence on X', so that $\Phi \in \mathfrak{Z}_t(X')$. The remainder has been established above. $\Box$

We shall see in a remark below Corollary 11.10 that, in the above circumstances, we can have $\mathfrak{Z}_t^{(1)}(A'') \cap X' \subsetneq \mathfrak{Z}_t(X')$.

COROLLARY 5.10. *Let A be a commutative Banach algebra such that A'' has a mixed identity. Then $\mathfrak{Z}(A'') \subset \mathfrak{Z}((A' \cdot A)')$.*

PROOF. Since A is commutative, we have

$$\mathfrak{Z}(A'') = \mathfrak{Z}_t^{(1)}(A'') = \mathfrak{Z}_t^{(2)}(A'') .$$

Set $X = A' \cdot A$. By Theorem 2.21, $\mathfrak{Z}(A'') \subset X'$, and so $\mathfrak{Z}(A'') \subset \mathfrak{Z}_t(X')$ by Proposition 5.9. $\Box$

The proof of the following result is similar to one first given in [L1, Theorem 1].

PROPOSITION 5.11. *Let A be a Banach algebra with a bounded left approximate identity, and let X be a left-introverted subspace of A'. Let $T \in \mathcal{B}(X)$. Then T is a left A-module homomorphism if and only if there exists $\Phi \in X'$ such that*

$$T\lambda = \lambda \cdot \Phi \quad (\lambda \in X) . \tag{5.4}$$

PROOF. Suppose that T has the form specified in equation (5.4). Then

$$T(a \cdot \lambda) = a \cdot \lambda \cdot \Phi = a \cdot T\lambda \quad (a \in A,\, \lambda \in X) ,$$

and so T is an A-module homomorphism.

Conversely, suppose that T is an A-module homomorphism. Let (e_α) be a bounded left approximate identity in A, and, as before, regard each e_α as an element of X'. The net $(T'e_\alpha)$ is bounded in X', and so has an accumulation point, say Φ, in $(X', \sigma(X', X))$; by passing to a subnet, we may suppose that $T'e_\alpha \to \Phi$. Now, for each $a \in A$ and $\lambda \in X$, we have

$$\begin{aligned}\langle a,\, T\lambda\rangle &= \lim_\alpha \langle e_\alpha a,\, T\lambda\rangle = \lim_\alpha \langle e_\alpha,\, a \cdot T\lambda\rangle \\ &= \lim_\alpha \langle e_\alpha,\, T(a \cdot \lambda)\rangle = \lim_\alpha \langle T'e_\alpha,\, a \cdot \lambda\rangle \\ &= \langle \Phi,\, a \cdot \lambda\rangle = \langle a,\, \lambda \cdot \Phi\rangle\,,\end{aligned}$$

and so $T\lambda = \lambda \cdot \Phi$, as required. □

There are two results from [LU] that we state for interest, but shall not use.

THEOREM 5.12. *Let A be a Banach algebra with a bounded approximate identity. Then $A \cdot \mathfrak{Z}_t^{(1)}(A'') = A \cdot \mathfrak{Z}_t((A' \cdot A)')$.*

PROOF. This is [LU, Corollary 3.2]. □

THEOREM 5.13. *Let A be a Banach algebra with a sequential bounded approximate identity. Suppose that A is weakly sequentially complete and that $A \cdot \mathfrak{Z}_t^{(1)}(A'') \subset A$. Then A is left strongly Arens irregular.*

PROOF. This is [LU, Theorem 3.4a]. □

Applications of the following theorem will appear within Example 6.2 and Theorem 7.25.

THEOREM 5.14. *Let A be a Banach algebra with an approximate identity of bound 1, and let X be a faithful, left-introverted submodule of A'. Then there is a continuous embedding*

$$\theta : (\mathcal{M}(A),\, \circ\,) \to (X',\, \Box\,)$$

such that

$$\text{(5.5)}\quad \langle \theta((L,R)), \lambda \cdot a\rangle = \langle Ra, \lambda\rangle \quad (a \in A,\, \lambda \in X,\, (L,R) \in \mathcal{M}(A))\,.$$

In the case where $X = \overline{XA}$, the range of θ is contained in $\mathfrak{Z}_t(X')$.

PROOF. The space A'' has a mixed identity, say Φ_0, with $\|\Phi_0\| = 1$, and so (2.16) describes an isometric embedding $\kappa : \mathcal{M}(A) \to (A'',\, \Box\,)$. Let $q : (A'',\, \Box\,) \to (X',\, \Box\,)$ be the quotient map. Then

$$\theta = q \circ \kappa : \mathcal{M}(A) \to (X',\, \Box\,)$$

is a continuous embedding. Let $a \in A$, $\lambda \in X$, and $(L,R) \in \mathcal{M}(A)$. Then $\lambda \cdot a \in X$ and

$$\langle R''(\Phi_0),\, \lambda \cdot a\rangle = \lim_\alpha \langle a \cdot Re_\alpha,\, \lambda\rangle = \lim_\alpha \langle R(ae_\alpha),\, \lambda\rangle = \langle Ra,\, \lambda\rangle\,,$$

where $(e_\alpha) \subset A_{[1]}$ is an approximate identity which converges to Φ_0. Hence (5.5) follows. Since X is faithful, θ is an injection.

Now suppose that $X = \overline{XA}$, so that we have $X = X \cdot A$. Let $(L, R) \in \mathcal{M}(A)$, and set $\Phi = \theta((L, R)) \in X'$. Let $\Psi_\nu \to \Psi$ in X'. Then, for each $a \in A$ and $\lambda \in X$, we have

$$\begin{aligned} \langle \Phi \,\Box\, \Psi_\nu,\, \lambda \cdot a\rangle &= \langle \Phi, \Psi_\nu \cdot \lambda \cdot a\rangle = \langle Ra, \Psi_\nu \cdot \lambda\rangle \\ &= \langle \Psi_\nu,\, \lambda \cdot Ra\rangle \to \langle \Psi,\, \lambda \cdot Ra\rangle \\ &= \langle Ra,\, \Psi \cdot \lambda\rangle = \langle \Phi \,\Box\, \Psi, \lambda \cdot a\rangle\,, \end{aligned}$$

and so $L_\Phi(\Psi_\nu) \to L_\Phi(\Psi)$. Thus L_Φ is continuous, and hence we have shown that $\Phi \in \mathfrak{Z}_t(X')$. □

In special cases that we shall consider later, a fixed space X may be a left-introverted subspace of two related dual spaces A' and B'. We investigate when the corresponding products $\Box_A$ and $\Box_B$ on X' coincide.

In the following result, we regard $\mathcal{M}(A)$ as a subalgebra of $(X', \Box)$ by using the embedding prescribed in the above theorem.

THEOREM 5.15. *Let A be a Banach algebra with an approximate identity of bound 1, and let X be a faithful, left-introverted submodule of A'. Let B be a $\|\cdot\|$-closed, unital subalgebra of $\mathcal{M}(A)$ such that B is $\sigma(X', X)$-dense in X'. Then X is a subspace of B' which is a faithful, left-introverted submodule. Further,*

$$\Phi \,\Box_A \Psi = \Phi \,\Box_B \Psi \quad (\Phi, \Psi \in X')\,. \tag{5.6}$$

PROOF. Fix $\lambda \in X$, and take S to be the $\sigma(A', A)$-closure of $A_{[1]} \cdot \lambda$. By Proposition 5.3, we have $S \subset X$.

For each $b \in B_{[1]} \subset X'$, there exists $\Phi \in A''_{[1]}$ such that $\Phi \mid X = b$. For each $a \in A$, we have

$$\langle a,\, \Phi \cdot \lambda\rangle = \langle \Phi,\, \lambda \cdot a\rangle = \langle b,\, \lambda \cdot a\rangle = \langle ab,\, \lambda\rangle = \langle a,\, b \cdot \lambda\rangle\,,$$

and so $\Phi \cdot \lambda = b \cdot \lambda$. Take a net (a_α) in $A_{[1]}$ such that $a_\alpha \to \Phi$ in the topology $\sigma(A'', A')$. Then $a_\alpha \cdot \lambda \to \Phi \cdot \lambda = b \cdot \lambda$ in $(A', \sigma(A', A))$. Thus $b \cdot \lambda \in S$.

Now take $\Gamma \in B''$ with $\|\Gamma\| = 1$. There is a net (b_β) in $B_{[1]}$ such that $b_\beta \to \Gamma$ in $(B'', \sigma(B'', B'))$, and then $b_\beta \cdot \lambda \cdot a \to \Gamma \cdot \lambda \cdot a$ in $(X, \sigma(B', B))$ for each $a \in A$. The identity multiplier belongs to B, and so it follows from (5.5) that

$$\langle a,\, b_\beta \cdot \lambda\rangle \to \langle a,\, \Gamma \cdot \lambda\rangle \quad (a \in A)\,.$$

Thus $b_\beta \cdot \lambda \to \Gamma \cdot \lambda$ in $(A', \sigma(A', A))$. Since $(b_\beta \cdot \lambda) \subset S$ and S is $\sigma(A', A)$-closed, it follows that $\Gamma \cdot \lambda \in S$, and so $\Gamma \cdot \lambda \in X$. This shows that X is left-introverted as a subspace of B'. Clearly X is faithful.

Take $\lambda \in X$ and $\Psi \in X'$. Then we can calculate $\Psi \cdot \lambda$ in both A' and B', say these elements are $\Psi \cdot_A \lambda$ and $\Psi \cdot_B \lambda$, respectively. We *claim* that they are equal as elements of X.

First, suppose that $\Psi = b \in B$. For each $a \in A$, we have

$$\langle a,\, b \cdot_A \lambda \rangle = \langle a,\, b \cdot_B \lambda \rangle = \langle a \,\Box\, b,\, \lambda \rangle\,,$$

where we are regarding a and b as elements of $\mathcal{M}(A)$ and hence of $(X', \Box\,)$. It follows that $b \cdot_A \lambda = b \cdot_B \lambda$.

Second, let Ψ be an arbitrary element of X'. By hypothesis, there is a net (b_β) in B such that $b_\beta \to \Psi$ in the topology $\sigma(X', X)$. Take $b \in B$. Then

$$\begin{aligned} \langle b,\, \Psi \cdot_B \lambda \rangle &= \langle \Psi,\, \lambda \cdot b \rangle = \lim_\beta \langle b_\beta,\, \lambda \cdot b \rangle \\ &= \lim_\beta \langle b,\, b_\beta \cdot_B \lambda \rangle = \lim_\beta \langle b,\, b_\beta \cdot_A \lambda \rangle \end{aligned}$$

by the earlier result. Now

$$\langle b,\, b_\beta \cdot_A \lambda \rangle = \langle b \,\Box_A b_\beta,\, \lambda \rangle \quad \text{and} \quad \lim_\beta \langle b \,\Box_A b_\beta, \lambda \rangle = \langle b \,\Box_A \Psi, \lambda \rangle$$

because $b \in \mathfrak{Z}_t(X')$. Hence $\langle b, \Psi \cdot_B \lambda \rangle = \langle b, \Psi \cdot_A \lambda \rangle$. Since B is $\sigma(X', X)$-dense in X', it follows that $\Psi \cdot_B \lambda = \Psi \cdot_A \lambda$.

Finally, take $\Phi, \Psi \in X'$. For each $\lambda \in X$, we have

$$\langle \Phi \,\Box_A \Psi, \lambda \rangle = \langle \Phi,\, \Psi \cdot_A \lambda \rangle = \langle \Phi,\, \Psi \cdot_B \lambda \rangle = \langle \Phi \,\Box_B \Psi, \lambda \rangle\,,$$

and this gives (5.6). $\Box$

CHAPTER 6

Banach Algebras of Operators

Before we begin our main study of Beurling algebras, we give some further examples. These examples are Banach algebras of operators on a Banach space E; for background on this topic, see [D, Chapter 2.5].

Let E and F be Banach spaces, and let $T \in \mathcal{B}(E, F)$. Then the *dual* T' of T is defined by

$$\langle x, T'\lambda \rangle = \langle Tx, \lambda \rangle \quad (x \in E,\ \lambda \in F'),$$

so that $T' \in \mathcal{B}(F', E')$. We have $\|T'\| = \|T\| \quad (T \in \mathcal{B}(E, F))$, and also $(S \circ T)' = T' \circ S' \ (S, T \in \mathcal{B}(E))$. We define

$$\mathcal{B}(E)^a = \{T' : T \in \mathcal{B}(E)\} \subset \mathcal{B}(E'),$$

so that $\mathcal{B}(E)^a$ is a closed, unital subalgebra of $\mathcal{B}(E')$ and $\mathcal{B}(E)^a$ consists of the operators $U \in \mathcal{B}(E')$ such that U is a continuous linear mapping from $(E', \sigma(E', E))$ into itself. Notice that $\mathcal{B}(E)^a = \mathcal{B}(E')$ if and only if E is reflexive.

The second dual of an operator $T \in \mathcal{B}(E, F)$ is $T'' : E'' \to F''$, and $T'' \circ \kappa_E = \kappa_F \circ T$.

Let E and F be Banach spaces. We identify $F \otimes E'$ with the space $\mathcal{F}(E, F)$ of continuous, finite-rank operators from E to F; indeed, for $y_0 \in F$ and $\lambda_0 \in E'$, the element $y_0 \otimes \lambda_0 \in F \otimes E'$ corresponds to the rank-one operator

$$y_0 \otimes \lambda_0 : x \mapsto \langle x,\, \lambda_0 \rangle y_0, \quad E \to F.$$

In particular $E \otimes E' = \mathcal{F}(E)$. Let $T = x_0 \otimes \lambda_0 \in E \otimes E'$. Then clearly $T' = \lambda_0 \otimes \kappa_E(x_0) \in E' \otimes E''$. The product in $E \otimes E'$ from $\mathcal{F}(E)$ is specified by

$$(x_1 \otimes \lambda_1) \circ (x_2 \otimes \lambda_2) = \langle x_2,\, \lambda_1 \rangle x_1 \otimes \lambda_2 \quad (x_1, x_2 \in E,\ \lambda_1, \lambda_2 \in E').$$

Let $x_0 \otimes \lambda_0 \in E \otimes E'$ and $T \in \mathcal{B}(E)$. Then

$$(6.1) \qquad T \circ (x_0 \otimes \lambda_0) = Tx_0 \otimes \lambda_0, \quad (x_0 \otimes \lambda_0) \circ T = x_0 \otimes T'\lambda_0.$$

Let E be a Banach space. We write $E \check{\otimes} E'$ for the injective tensor product of E with E'; this space is identified with $\mathcal{A}(E) = \overline{\mathcal{F}(E)}$, the closed ideal of *approximable operators* in $\mathcal{B}(E)$. We also write $E \widehat{\otimes} E'$ for the projective tensor product of E with E', with projective norm

$\|\cdot\|_\pi$. In this way $(E\widehat{\otimes}E', \|\cdot\|_\pi)$ is a Banach algebra, which is called the *nuclear algebra* of E in [D, Definition 2.5.4]. We write $\mathcal{N}(E)$ for the ideal of nuclear operators on E, as in [D, Chapter 2.5]; however, we now write $\|\cdot\|_{\mathcal{N}}$ for the nuclear norm on $\mathcal{N}(E)$.

Let E be a Banach space, and set $\mathfrak{A} = K(E)$. Then $\mathfrak{A}$ has a bounded left approximate identity if and only if E has the bounded compact approximation property ([Dix], [D, Theorem 2.9.37]). Now suppose that E' has the bounded approximation property (BAP). Then E itself has the BAP, $\mathfrak{A} := \mathcal{A}(E) = \mathcal{K}(E)$, and $\mathfrak{A}$ has a bounded approximate identity (see [D, Theorem 2.9.37]), so that $\mathfrak{A}''$ has a mixed identity.

Let E be a Banach space, and suppose that $\mathfrak{A}$ is a subalgebra of $\mathcal{B}(E)$ such that $\mathfrak{A}$ contains the finite-rank operators and $\mathfrak{A}$ is a Banach algebra with respect to some norm, so that $\mathfrak{A}$ is a *Banach operator algebra* in the sense of [D, Definition 2.5.1]. In fact the embedding of $\mathfrak{A}$ into $\mathcal{B}(E)$ is necessarily continuous.

First, suppose that $\mathfrak{A}$ is Arens regular. Then E is necessarily reflexive. Indeed, assume towards a contradiction that E is a non-reflexive Banach space. Then there are sequences (x_m) in $E_{[1]}$ and (λ_n) in $E'_{[1]}$ such that $(\langle x_m, \lambda_n\rangle)$ has unequal repeated limits. Take $\Lambda \in \mathfrak{A}'$ such that $\Lambda \mid \mathcal{A}(E) \neq 0$. Then there exist elements $x \in E$ and $\lambda \in E'$ such that $\langle x \otimes \lambda, \Lambda\rangle = 1$. Set $S_m = x_m \otimes \lambda$ $(m \in \mathbb{N})$ and $T_n = \lambda_n \otimes x$ $(n \in \mathbb{N})$, so that (S_m) and (T_n) are sequences in $\mathfrak{A}$. Then $T_n S_m = \langle x_m, \lambda_n\rangle x \otimes \lambda$ $(m, n \in \mathbb{N})$, and so $(\langle T_n S_m, \Lambda\rangle)$ has unequal repeated limits. Thus Λ is not weakly almost periodic, and so $WAP(\mathfrak{A}) \neq \mathfrak{A}'$. (This argument, from [Y3, Theorem 3], shows that $WAP(\mathcal{A}(E)) = \{0\}$.) In fact, let E be a non-reflexive Banach space such that E' has BAP. Then $\mathfrak{A} = \mathcal{A}(E)$ has a bounded approximate identity, but

$$\{0\} = WAP(\mathfrak{A}) \subsetneq \mathfrak{A}' \cdot \mathfrak{A},$$

and so we do not always have equality in the inclusion of Proposition 3.12. (A related result is given as Proposition 3.3 of [DuU]: for each infinite-dimensional Banach space E with the approximation property, we have $AP(\mathcal{K}(E)) = \{0\}$.)

In the other direction, in the case where E is reflexive, the Banach algebras $E\,\widehat{\otimes}\,E'$, $\mathcal{N}(E)$, $\mathcal{A}(E)$, and $\mathcal{K}(E)$ are all Arens regular. For these and other related results, see [D, Theorem 2.6.23]. The results are due to Young [Y3] and Ülger [U3]; see also [PyU].

It was left open in [D] and elsewhere whether or not $\mathcal{B}(E)$ is Arens regular in various cases where E is reflexive; however, specific reflexive spaces E for which $\mathcal{B}(E)$ is not Arens regular have been given in

[Y3] and [PyU]. In fact, it is shown in [Y3, Theorem 4] that, for each locally compact group G, there is a reflexive Banach space E and an isometric isomorphism from $L^1(G)$ onto a closed subalgebra of $\mathcal{B}(E)$. Since $L^1(G)$ is not Arens regular (for each infinite group G), it cannot be that $\mathcal{B}(E)$ is Arens regular. Let E be a reflexive Banach space such that $\mathcal{B}(E)$ is not Arens regular, and set $\mathfrak{A} = \mathcal{K}(E)$. Then $\mathfrak{A}$ is Arens regular, but $\mathfrak{A}''$ is not.

We now sketch an important advance on this question due to Daws; for a different, more abstract, presentation, see [Da1] and [Da2].

Let E be a Banach space, and set $X = E \widehat{\otimes} E'$. For $\mu \in X'$, define $T_\mu : E \to E''$ by

$$\langle T_\mu x,\, \lambda\rangle = \langle x \otimes \lambda,\, \mu\rangle \quad (x \in E,\, \lambda \in E')\,.$$

Then $\mu \mapsto T_\mu,\ X' \to \mathcal{B}(E, E'')$, is an isometric linear bijection.

Now suppose that E is reflexive. Then we have $X' = \mathcal{B}(E)$, and X is a $\mathcal{B}(E)$-submodule of $\mathcal{B}(E)'$, and so $\mathcal{B}(E)$ is a dual Banach algebra in the sense of Definition 2.6. Let $\Lambda \in \mathcal{B}(E)' = X''$: we seek to prove that Λ is weakly almost periodic. By Theorem 3.14, this is sufficient to establish that $\mathcal{B}(E)$ is Arens regular.

First we introduce some more Banach spaces. Let E be a Banach space, and set

$$\ell^2(E) = \left\{ x = (x_k) \in E^{\mathbb{N}} : \|x\| = \left(\sum_{k=1}^{\infty} \|x_k\|^2\right)^{1/2} < \infty \right\},$$

so that $\ell^2(E)$ is a Banach space. The dual space of $\ell^2(E)$ is $\ell^2(E')$, with the duality

$$\langle x,\, \lambda\rangle = \sum_{k=1}^{\infty} \langle x_k,\, \lambda_k\rangle \quad (x = (x_k) \in \ell^2(E),\, \lambda = (\lambda_k) \in \ell^2(E'))\,.$$

In the case where E is reflexive, $\ell^2(E)$ is also reflexive.

For $S \in \mathcal{B}(E)$, define

$$\widetilde{S}x = (Sx_k) \quad (x = (x_k) \in \ell^2(E))\,.$$

Then $\widetilde{S} \in \mathcal{B}(\ell^2(E))$, and the dual $\widetilde{S}'$ of $\widetilde{S}$ is specified by $\widetilde{S}'\lambda = (S'\lambda_k)$ for $\lambda = (\lambda_k) \in \ell^2(E')$. The map $S \mapsto \widetilde{S},\ \mathcal{B}(E) \to \mathcal{B}(\ell^2(E))$, is a bounded linear operator.

Next we recall the definition of the ultrapower of a Banach space; see [Hei] for a clear account of the basic results in this area. Let F be a Banach space, and let I be a directed index set, and consider

$$\ell^\infty(F, I) = \{y = (y_\alpha : \alpha \in I) : \|y\| = \sup_{\alpha \in I} \|y_\alpha\| < \infty\}\,,$$

again a Banach space. Finally, let $\mathcal{U}$ be an ultrafilter on I dominating the order filter, and set $y \sim 0$ in $\ell^\infty(F, I)$ if, for each $\varepsilon > 0$, the set

$$\{\alpha \in I : \|y_\alpha\| < \varepsilon\}$$

belongs to $\mathcal{U}$. Then $\{y \in \ell^\infty(F, I) : y \sim 0\}$ is a closed subspace of $\ell^\infty(F, I)$, and the quotient space $\ell^\infty(F, I)/\!\sim$ is a Banach space denoted by $F_\mathcal{U}$; indeed, $F_\mathcal{U}$ is the *ultrapower* of F in the category of Banach spaces. The coset of $y \in \ell^\infty(F, I)$ is also denoted by y in $F_\mathcal{U}$. For each $T \in \mathcal{B}(F)$, the map

$$T_\mathcal{U} : (y_\alpha) \mapsto (Ty_\alpha), \quad F_\mathcal{U} \to F_\mathcal{U}\,,$$

is a bounded linear operator with $\|T_\mathcal{U}\| = \|T\|$, called the *ultrapower* of T (see [Hei, Chapter 2]). The map $T \mapsto T_\mathcal{U}$, $\mathcal{B}(F) \to \mathcal{B}(F_\mathcal{U})$, is an isometric embedding. By the first paragraph of [Hei, Proposition 7.1], there is an isometric embedding of $(F')_\mathcal{U}$ into $(F_\mathcal{U})'$: for $(y_\alpha) \in F_\mathcal{U}$ and $(\lambda_\alpha) \in (F')_\mathcal{U}$, the duality is given by

$$\langle (y_\alpha), (\lambda_\alpha)\rangle = \lim_{\alpha\in\mathcal{U}} \langle y_\alpha, \lambda_\alpha\rangle\,,$$

where this limit always exists.

Let E be a Banach space. The Banach space E is *super-reflexive* if there is an equivalent norm $|||\cdot|||$ on E such that $(E, |||\cdot|||)$ is uniformly convex (and we suppose in this case that the original norm $\|\cdot\|$ is uniformly convex). Let $\mathcal{U}$ be an ultrafilter on an index set I. Then $E_\mathcal{U}$ is a reflexive Banach space whenever E is super-reflexive; in fact, by [Hei, Proposition 6.4], a Banach space E is super-reflexive if and only if each ultrapower $E_\mathcal{U}$ is reflexive. For example, all Banach spaces of the form $L^p(\mu)$ for a positive measure μ, where $1 < p < \infty$, are uniformly convex, and hence super-reflexive.

Finally, start with a uniformly convex space E, set $F = \ell^2(E)$, and let $\mathcal{U}$ be an ultrafilter on an index set. Then F is also uniformly convex, and so $F_\mathcal{U}$ is reflexive.

We now return to $\Lambda \in \mathcal{B}(E)' = X''$. There is a directed set I and a net $(z_\alpha : \alpha \in I)$ in X such that $\|z_\alpha\| \le \|\Lambda\|$ $(\alpha \in I)$ and $z_\alpha \to \Lambda$ in

$$\lim_{\alpha\in I} \langle z_\alpha, T\rangle = \langle T, \Lambda\rangle \quad (T \in \mathcal{B}(E))\,.$$

Fix $\alpha \in I$. We can represent z_α as

$$z_\alpha = \sum_{k=1}^{\infty} x_{k,\alpha} \otimes \mu_{k,\alpha}\,,$$

where $x_\alpha := (x_{k,\alpha} : k \in \mathbb{N}) \in E^{\mathbb{N}}$ and $\mu_\alpha := (\mu_{k,\alpha} : k \in \mathbb{N}) \in (E')^{\mathbb{N}}$, and we have

$$\sum_{k=1}^{\infty} \|x_{k,\alpha}\| \, \|\mu_{k,\alpha}\| \leq m \,,$$

where $m = \|\Lambda\| + 1$. Set $F = \ell^2(E)$. Then, in fact, we can suppose that $\|x_{k,\alpha}\| = \|\mu_{k,\alpha}\|$ $(k \in \mathbb{N})$, and so $x_\alpha \in F$ and $\mu_\alpha \in \ell^2(E') = F'$, with $\|x_\alpha\|^2 \leq m$ and $\|\mu_\alpha\|^2 \leq m$.

Let $\mathcal{U}$ be an ultrafilter on the index set I such that $\mathcal{U}$ dominates the order filter. Define the maps

$$\begin{aligned} U : S \mapsto (\widetilde{S}'\mu_\alpha : \alpha \in I), \qquad & \mathcal{B}(E) \to (F_{\mathcal{U}})' \,, \\ V : S \mapsto (\widetilde{S}x_\alpha \alpha \in I), \qquad & \mathcal{B}(E) \to F_{\mathcal{U}} \,. \end{aligned}$$

Then it is easy to check that each of U and V is a bounded linear operator.

Let $S, T \in \mathcal{B}(E)$. For each $k \in \mathbb{N}$ and $\alpha \in I$, we have

$$\langle x_{k,\alpha} \otimes \mu_{k,\alpha}, \, ST \rangle = \langle STx_{k,\alpha}, \, \mu_{k,\alpha} \rangle = \langle Tx_{k,\alpha}, \, S'\mu_{k,\alpha} \rangle \,,$$

and so, for each $\alpha \in I$, we have

$$\langle z_\alpha, \, ST \rangle = \sum_{k=1}^{\infty} \langle x_{k,\alpha} \otimes \mu_{k,\alpha}, \, ST \rangle = \sum_{k=1}^{\infty} \langle Tx_{k,\alpha}, \, S'\mu_{k,\alpha} \rangle = \langle \widetilde{T}x_\alpha, \, \widetilde{S}'\mu_\alpha \rangle \,.$$

Thus we see that

$$\langle ST, \, \Lambda \rangle = \lim_{\alpha \in I} \langle z_\alpha, \, ST \rangle = \lim_{\alpha \in I} \left\langle \widetilde{T}x_\alpha, \, \widetilde{S}'\mu_\alpha \right\rangle \,.$$

But

$$\langle V(T), \, U(S) \rangle = \lim_{\alpha \in \mathcal{U}} \left\langle \widetilde{T}x_\alpha, \, \widetilde{S}'\mu_\alpha \right\rangle \,,$$

and the ultrafilter $\mathcal{U}$ dominates the order filter on I, and so we finally see that $\langle ST, \, \Lambda \rangle = \langle V(T), \, U(S) \rangle$.

It follows from Proposition 3.13 that the continuous linear functional Λ is weakly almost periodic whenever the space $F_{\mathcal{U}}$ is reflexive; we have explained that this is always the case when the original space E is super-reflexive. Thus the following theorem of Daws answers a question specifically raised in [Y3, p. 109] and [DuH].

THEOREM 6.1. *Let E be a super-reflexive Banach space. Then $\mathcal{B}(E)$ is Arens regular.* □

In the case where H is a Hilbert space, $(\mathcal{B}(H)'', \Box\,)$ is a C^*-algebra, and hence is semisimple. There are examples of Banach spaces E for which $(\mathcal{B}(E)'', \Box\,)$ is not semisimple. We suspect, but cannot prove, that $(\mathcal{B}(\ell^p(\mathbb{N}))'', \Box\,)$ is semisimple whenever $1 < p < \infty$.

[*Added in August, 2004:* In fact it has been proved by Daws and Read [DaRe] that, for $1 < p < \infty$, the Banach algebra $(\mathcal{B}(\ell^p(\mathbb{N}))'', \Box)$ is semisimple if and only if $p = 2$; it is now conceivable that $(\mathcal{B}(E)'', \Box)$ is semisimple (for a reflexive Banach space E) if and only if E is linearly homeomorphic to a Hilbert space. For further details, see [Da2].]

Before giving the next example, we make some remarks.

First, let $(A, \cdot)$ be an algebra, and let $P : A \to A$ be a projection on A, so that P is a linear map such that $P^2 = P$. Set $B = P(A)$ and $J = \ker P$; clearly, $A = B \oplus J$ as linear spaces. Now suppose, further, that:

(1) B is a subalgebra of A;

(2) J is a left ideal in A.

We define a bilinear map

$$(a, b) \mapsto a \times b := P(a)b, \quad A \times A \to A\,.$$

Since (1) and (2) hold, the product in $B \oplus J$ satisfies the equation

$$(b_1, 0) \cdot (b_2, x) = (b_1 b_2, b_1 x) \quad (b_1, b_2 \in B,\ x \in J)\,,$$

and so $P(Pa \cdot b) = Pa \cdot Pb$ $(a, b \in A)$. It follows that

$$(a \times b) \times c = a \times (b \times c) \quad (a, b, c \in A)\,,$$

and thus $\times$ is an (associative) product on A. Further,

$$P(a \times b) = Pa \cdot Pb = Pa \times Pb \quad (a, b \in A)\,,$$

and so P is a epimorphism from $(A, \times)$ onto the subalgebra $(B, \times)$, and hence $(A, \times) = B \ltimes J$.

Let R denote the radical of $(A, \times)$. Since

$$a \times b = 0 \quad (a \in J,\ b \in A)\,,$$

we have $J \subset R$.

We make the following further assumption:

(3) B is a semisimple algebra.

Then $J = R$, and hence $(A, \times)$ is decomposable.

For our further discussion in this chpater, let E be a Banach space, with second dual E''. For each $U \in \mathcal{B}(E'')$, set

$$\eta(U) = \kappa_E' \circ U' \circ \kappa_{E'}, \quad E' \to E'\,,$$

so that $\eta(U) \in \mathcal{B}(E')$, and then set

$$\mathcal{Q}(U) = \eta(U)'\,,$$

so that $\mathcal{Q}(U) \in \mathcal{B}(E'')$ and $\mathcal{Q}$ is a bounded linear operator on $\mathcal{B}(E'')$. Alternatively, we see that $\mathcal{Q}(U)$ is defined from U by first restricting

U to E, and then extending this latter operator by weak-$*$ continuity in both its domain and range spaces. From both descriptions of $\mathcal{Q}$, it is clear that $\mathcal{Q}$ is a projection on $\mathcal{B}(E'')$ with $\|\mathcal{Q}\| = 1$ and that the range of $\mathcal{Q}$ is exactly $\mathcal{B}(E')^a$. In particular, $\mathcal{Q}(I_{E''}) = I_{E''}$. Let $\mathfrak{J}$ be the kernel of $\mathcal{Q}$, so that $\mathfrak{J}$ is the set of elements $U \in \mathcal{B}(E'')$ such that $U \mid E = 0$.

Let E be a Banach space, and again consider the space $X = E\widehat{\otimes}E'$. For each $\mu \in X'$, define $T_\mu : E' \to E'$ by

$$\langle x, T_\mu\lambda\rangle = \langle x \otimes \lambda, \mu\rangle \quad (x \in E, \lambda \in E'). \tag{6.2}$$

Then $T_\mu \in \mathcal{B}(E')$, and the map $\mu \mapsto T_\mu, \ X' \to \mathcal{B}(E')$, is an isometric isomorphism. Thus we have the identification $(E\widehat{\otimes}E')' = \mathcal{B}(E')$.

Let E be a Banach space with the approximation property (AP), so that $\mathcal{A}(E) = \mathcal{K}(E)$, and set $\mathfrak{A} = \mathcal{K}(E)$. It is of interest to calculate the two Arens products on $\mathfrak{A}''$.

Let E be a Banach space. Each $T \in \mathcal{B}(E)$ defines a linear functional on $E \otimes E'$ by the action

$$S = \sum_{j=1}^{n} x_j \otimes \lambda_j \mapsto \sum_{j=1}^{n} \langle x_j, T'\lambda_j\rangle = \mathrm{tr}\,(ST)\,,$$

where 'tr' denotes the trace. In the case where this linear functional is continuous on $(\mathcal{F}(E), \|\cdot\|)$, the operator T is an *integral operator*, and its continuous extensions to $\mathcal{A}(E)$ is denoted by $\check{T}$, with norm $\|\check{T}\|_{\mathcal{I}}$. The set of all integral operators on E is denoted by $\mathcal{I}(E)$; it is clear that $(\mathcal{I}(E), \|\cdot\|_{\mathcal{I}})$ is a Banach operator ideal in the sense of [D, Definition 2.5.1]. Each nuclear operator on E is an integral operator, and we have $\|T\|_{\mathcal{I}} \leq \|T\|_{\mathcal{N}}$ $(T \in \mathcal{N}(E))$. It is shown in [DiU, Theorem VIII.3.8] that, in the case where E has the metric approximation property, we have

$$\|T\|_{\mathcal{I}} = \|T\|_{\mathcal{N}} \quad (T \in \mathcal{N}(E))\,,$$

and so $\mathcal{N}(E)$ is a closed ideal in $\mathcal{I}(E)$. See also [DeF, p. 193].

Let $\mu : E \otimes E' \to \mathbb{C}$ be a linear functional, and define T_μ on E' as in (6.2). Then μ gives a continuous linear functional on $E\check{\otimes}E'$ if and only if $T_\mu \in \mathcal{I}(E')$ [DiU, Corollary VIII.2.12], and so the map $\mu \mapsto T_\mu$ is an isometric linear isomorphism from $(E\check{\otimes}E')' = \mathcal{A}(E)'$ into $(\mathcal{I}(E'), \|\cdot\|_{\mathcal{I}})$. For details, see [Pa1, Chapter 1.7.12] and [Da2].

Now suppose that E is a Banach space such that E' has AP and the Radon–Nikodým property (RNP). (For example, in the case where we set $E = c_0$, the dual space $E' = \ell^1$ has AP and RNP. For a discussion of RNP, see [DiU].) Then it is a theorem of Grothendieck that

$$\mathcal{I}(E') = \mathcal{N}(E') = E'\widehat{\otimes}E''\,.$$

(For a closely related result, see the references [DeF, 16.5] and [DiU, Theorem VIII.4.6].) Again set $\mathfrak{A} = \mathcal{K}(E)$. Then we have:

$$\mathfrak{A} = E\check{\otimes}E'; \quad \mathfrak{A}' = E'\widehat{\otimes}E'' = \mathcal{N}(E');$$
$$\mathfrak{A}'' = (E'\widehat{\otimes}E'')' = \mathcal{B}(E'').$$

The embedding $\kappa_{\mathfrak{A}}$ of $\mathfrak{A}$ into $\mathfrak{A}''$ is the map $\kappa_{\mathfrak{A}} : A \mapsto A''$. The duality between $\mathfrak{A}$ and $\mathfrak{A}'$ is specified by

$$\langle A,\, \mu \otimes \Lambda\rangle = \langle \Lambda,\, A'\mu\rangle = \langle A''(\Lambda),\, \mu\rangle \tag{6.3}$$

for $A \in \mathfrak{A}$ and $\mu \in E'$, $\Lambda \in E''$, and the duality between $\mathfrak{A}'$ and $\mathfrak{A}''$ is specified by

$$\langle U,\, \mu \otimes \Lambda\rangle = \langle U(\Lambda),\, \mu\rangle \tag{6.4}$$

for $U \in \mathfrak{A}''$ and $\mu \in E'$, $\Lambda \in E''$. For a full description of these dualities, see [DeF, Chapter 16.7], for example.

We now describe the canonical module actions of $\mathfrak{A}$ on $\mathfrak{A}'$ and on $\mathfrak{A}''$. Thus, take $A \in \mathfrak{A}$, $T \in \mathfrak{A}' = \mathcal{N}(E')$, and $U \in \mathfrak{A}'' = \mathcal{B}(E'')$. Then:

$$\left.\begin{array}{lllllll} A \cdot T & = & T \circ A'; & T \cdot A & = & A' \circ T; \\ A \cdot U & = & A'' \circ U; & U \cdot A & = & U \circ A''. \end{array}\right\} \tag{6.5}$$

It is of interest to note that

$$\mathfrak{A}' = \mathfrak{A}' \cdot \mathfrak{A}, \quad \text{but that} \quad \mathfrak{A}' \neq \mathfrak{A} \cdot \mathfrak{A}'; \tag{6.6}$$

this is proved in [LU, Example 2.5]. It follows as in [LU, Proposition 2.10] that $\mathfrak{Z}_t^{(1)}(\mathfrak{A}'') \neq \mathfrak{Z}_t^{(2)}(\mathfrak{A}'')$, but we shall show more than this.

We shall now calculate the two Arens products on $\mathfrak{A}'' = \mathcal{B}(E'')$, still in the case where E' has AP and RNP. This calculation was first made by Palmer in [Pa1] and, for more general Banach spaces and in full detail, by M. Grosser in [G2]. For variety, we make the calculations in a slightly different way.

Let $U, V \in \mathfrak{A}''$, and take (A_α) and (B_β) to be nets in $\mathfrak{A}$ such that $\lim_\alpha A_\alpha = U$ and $\lim_\beta B_\beta = V$ (in the topology $\sigma(\mathfrak{A}'', \mathfrak{A}')$). Take $\mu \in E'$ and $\Lambda \in E''$. Then

$$\langle (A_\alpha B_\beta)''(\Lambda),\, \mu\rangle = \langle B_\beta''(\Lambda),\, A_\alpha'(\mu)\rangle .$$

For each α, we have $\lim_\beta \langle B_\beta'',\, A_\alpha'\mu \otimes \Lambda\rangle = \langle V,\, A_\alpha'\mu \otimes \Lambda\rangle$, and so

$$\lim_\beta \langle B_\beta''(\Lambda),\, A_\alpha'(\mu)\rangle = \langle V(\Lambda),\, A_\alpha'(\mu)\rangle = \langle A_\alpha''(V(\Lambda)),\, \mu\rangle .$$

Hence

$$\lim_\alpha \lim_\beta \langle (A_\alpha B_\beta)''(\Lambda),\, \mu\rangle = \langle U(V(\Lambda)),\, \mu\rangle .$$

This shows that

$$U \mathbin{\Box} V = \lim_{\alpha}\lim_{\beta} A_\alpha B_\beta = U \circ V \quad (U, V \in \mathcal{B}(E'')) . \tag{6.7}$$

As a preliminary to the calculation of $U \diamond V$, we make the following remark. Take $A = x \otimes \lambda \in E \otimes E' \subset \mathfrak{A}$, $T \in \mathfrak{A}'$, and $U \in \mathfrak{A}'' = \mathcal{B}(E'')$. Then

$$\begin{aligned} \langle A,\, T \cdot U\rangle &= \langle U,\, A \cdot T\rangle && \text{by (2.7)}\\ &= \langle U,\, T \circ A'\rangle && \text{by (6.5)}\\ &= \langle U,\, T\lambda \otimes \kappa_E(x)\rangle && \text{by (6.1)}\\ &= \langle (U \circ \kappa_E)(x), T\lambda\rangle && \text{by (6.4)}\\ &= \langle x,\, (\eta(U) \circ T)(\lambda)\rangle && \\ &= \langle A,\, \eta(U) \circ T\rangle && \text{by (6.2)}, \end{aligned}$$

and so $T \cdot U = \eta(U) \circ T$ in $\mathfrak{A}'$. (Similarly, $U \cdot T = \eta(U \circ T')$.)

We can now calculate $U \diamond V$ for $U, V \in \mathfrak{A}''$. First, we take an element $T = \mu \otimes \Lambda \in E' \otimes E''$. Then

$$\begin{aligned} \langle U \diamond V,\, T\rangle &= \langle V,\, T \cdot U\rangle && \text{by (2.8)}\\ &= \langle V,\, \eta(U) \circ (\mu \otimes \Lambda)\rangle && \\ &= \langle V,\, \eta(U)(\mu) \otimes \Lambda\rangle && \text{by (6.1)}\\ &= \langle V(\Lambda),\, \eta(U)(\mu)\rangle && \text{by (6.4)}\\ &= \langle \mathcal{Q}(U)(V(\Lambda)),\, \mu\rangle && \\ &= \langle \mathcal{Q}(U) \circ V,\, \mu \otimes \Lambda\rangle && \text{by (6.4)}\\ &= \langle \mathcal{Q}(U) \circ V,\, T\rangle . && \end{aligned}$$

This shows that

$$U \diamond V = \mathcal{Q}(U) \circ V \quad (U, V \in \mathcal{B}(E'')) . \tag{6.8}$$

In particular, the algebra $(\mathcal{B}(E)'', \diamond)$ is not unital.

The following example is an elaboration of Example 2.5 of [LU].

Example 6.2. Let E be a Banach space such that E' has AP and RNP and E is not reflexive. For example, we can take $E = c_0$. Set $\mathfrak{A} = \mathcal{K}(E)$. It is clear from (6.7) and (6.8) that

$$\mathfrak{Z}_t^{(1)}(\mathfrak{A}'') = \{U \in \mathcal{B}(E'') : U \circ V = \mathcal{Q}(U) \circ V \ \ (V \in \mathcal{B}(E''))\} ,$$

and so

$$\mathfrak{Z}_t^{(1)}(\mathfrak{A}'') = \{U \in \mathcal{B}(E'') : U = \mathcal{Q}(U)\} = \mathcal{B}(E')^a .$$

Similarly, we see that

$$\mathfrak{Z}_t^{(2)}(\mathfrak{A}'') = \{U \in \mathcal{B}(E'') : V \circ U = \mathcal{Q}(V) \circ U \quad (V \in \mathcal{B}(E''))\} .$$

For each $\Lambda \in E'' \setminus E$, there exists $V \in \mathcal{B}(E'')$ such that $V \mid E = 0$ and $V(\Lambda) = \Lambda$. It follows that

$$\mathfrak{Z}_t^{(2)}(\mathfrak{A}) = \{U \in \mathcal{B}(E'') : U(E'') \subset E\} .$$

(The identification of $\mathfrak{Z}_t^{(2)}(\mathcal{K}(c_0))$ in [LU, Chapter 6, j)] is not correct.)

Let $I_{E''}$ be the identity operator on E''. Then

$$I_{E''} \in \mathfrak{Z}_t^{(1)}(\mathfrak{A}) \setminus \mathfrak{Z}_t^{(2)}(\mathfrak{A}) .$$

On the other hand, take $x_0 \in E \setminus \{0\}$ and $\Gamma \in E'''$ with $\Gamma \mid E = 0$ and $\Gamma \neq 0$, and then define

$$U(\Lambda) = \langle \Lambda, \Gamma \rangle \, x_0 \quad (\Lambda \in E'') .$$

Then $U \in \mathcal{B}(E'')$ and $U(E'') \subset \mathbb{C}x_0 \subset E$, but $U \mid E = 0$, and so we have $\mathcal{Q}(U) = 0$, whence $\mathcal{Q}(U) \neq U$. Thus $U \in \mathfrak{Z}_t^{(2)}(\mathfrak{A}'') \setminus \mathfrak{Z}_t^{(1)}(\mathfrak{A}'')$. We conclude that

$$\mathfrak{Z}_t^{(1)}(\mathfrak{A}'') \not\subset \mathfrak{Z}_t^{(2)}(\mathfrak{A}'') \quad \text{and} \quad \mathfrak{Z}_t^{(2)}(\mathfrak{A}'') \not\subset \mathfrak{Z}_t^{(1)}(\mathfrak{A}'') ,$$

and so the two topological centres of $\mathfrak{A}$ are different. In particular,

$$\mathfrak{A} \subsetneq \mathfrak{Z}_t^{(1)}(\mathfrak{A}'') \subsetneq \mathfrak{A}'' \quad \text{and} \quad \mathfrak{A} \subsetneq \mathfrak{Z}_t^{(2)}(\mathfrak{A}'') \subsetneq \mathfrak{A}'' ,$$

and so $\mathfrak{A}$ is neither Arens regular nor either left or right strongly Arens irregular.

Let us consider how the above remarks relate to Theorem 5.14. Now $\mathcal{M}(\mathfrak{A}) = \mathcal{B}(E)$. We set $X = \mathfrak{A}'$. The algebra $(\mathfrak{A}'', \Box) = (\mathcal{B}(E''), \circ)$ is unital, and so it follows from [LU, Proposition 2.2a] that $\mathfrak{A}' = \mathfrak{A}' \cdot \mathfrak{A}$. Thus Theorem 5.14 applies to show that

$$\mathcal{B}(E)^{aa} \subset \mathfrak{Z}_t^{(1)}(\mathfrak{A}'') ;$$

in fact, we have established the stronger result that $\mathcal{B}(E')^a = \mathfrak{Z}_t^{(1)}(\mathfrak{A}'')$.

Temporarily set $Z = \mathfrak{Z}_t^{(1)}(\mathfrak{A}'') \cap \mathfrak{Z}_t^{(2)}(\mathfrak{A}'')$. We seek to identify Z.

First take $A \in \mathcal{W}(E)$. Then $A''(E'') \subset E$ and so $A'' \in \mathfrak{Z}_t^{(2)}(\mathfrak{A}'')$. Clearly $A'' \in \mathfrak{Z}_t^{(1)}(\mathfrak{A}'')$, and so we have shown that $A \in Z$. Hence $\mathcal{W}(E)^{aa} \subset Z$.

Now take $U \in Z$. Since $U \in \mathcal{B}(E')^a$, the operator U is continuous on the space $(E'', \sigma(E'', E'))$. Set $S := (E'')_{[1]}$, so that S is $\sigma(E'', E')$-compact. Then $U(S)$ is $\sigma(E'', E')$-compact. However $U(S) \subset E$ because $U \in \mathfrak{Z}_t^{(2)}(\mathfrak{A})$, and so $U(S)$ is $\sigma(E, E')$-compact in E, and hence $\sigma(E'', E''')$-compact in E''. This shows that $U \in \mathcal{W}(E'')$. Set $U = T'$, where $T \in \mathcal{B}(E')$. By Gantmacher's theorem [DfS, Theorem VI.4.8], we have $T \in W(E')$.

We now *claim* that $T \in \mathcal{B}(E)^a$. By [DfS, Exercise VI.9.13], we require T to be $\sigma(E', E)$-continuous on E'. To show that this is the case, it suffices, by [DfS, Theorem V.5.6], to show that $T\lambda_\alpha \to T\lambda$ in $(E', \sigma(E', E))$ whenever $\lambda_\alpha \to \lambda$ in $((E')_{[1]}, \sigma(E', E))$. Since T is weakly compact, we may suppose, by passing to a subnet, that $(T\lambda_\alpha)$ is convergent in $(E', \sigma(E', E))$, say $T\lambda_\alpha \to \mu \in E'$. Let $x \in E$, so that $\langle x, T\lambda_\alpha\rangle \to \langle x, \mu\rangle$. Set $y = T'x \in E''$: in fact, $y \in E$ because $T'(E'') \subset E$. Thus

$$\langle x,\, T\lambda_\alpha\rangle = \langle y,\, \lambda_\alpha\rangle \to \langle y,\, \lambda\rangle = \langle x,\, T\lambda\rangle\,,$$

and so $T\lambda = \mu$. Thus the claim is established.

Again by Gantmacher's theorem, we see that $T \in W(E)^a$, and this proves that $U \in W(E)^{aa}$.

We conclude that

$$\mathcal{W}(E)^{aa} = \left(\mathfrak{Z}_t^{(1)}(\mathfrak{A}'') \cap \mathfrak{Z}_t^{(2)}(\mathfrak{A}'')\right). \tag{6.9}$$

We now take special cases for the space E.

First take $E = c_0$. It is standard that $\mathcal{K}(c_0) = \mathcal{W}(c_0)$. To see this, take $T \in \mathcal{W}(c_0)$. Then $T' \in \mathcal{W}(\ell^1)$. Let (λ_n) be a bounded sequence in ℓ^1, so that $(T'\lambda_n)$ is also a bounded sequence in ℓ^1, and hence has a weakly convergent subsequence. But ℓ^1 has 'Schur's property': each weakly convergent sequence in ℓ^1 is norm-convergent [Co, V.5.2]. Thus T' is compact, and so T is compact. This gives us an example $\mathfrak{A} = \mathcal{K}(c_0)$ such that

$$\kappa_{\mathfrak{A}}(\mathfrak{A}) = \left(\mathfrak{Z}_t^{(1)}(\mathfrak{A}'') \cap \mathfrak{Z}_t^{(2)}(\mathfrak{A}'')\right). \tag{6.10}$$

Second, take $E = J$, the James space, so that J''/J has dimension 1. In particular, J''/J is a separable space, and so J' has RNP [DiU, p. 219]. Also J has AP, and so J fits into our present scenario. It also follows easily that $\mathcal{W}(J)$ is a closed ideal of codimension 1 in $\mathcal{B}(J)$ (and $\mathcal{W}(J)$ is the kernel of a character on $\mathcal{B}(J)$). However $\mathcal{K}(J)$ has infinite codimension in $\mathcal{B}(J)$, and so $\mathcal{K}(J) \neq \mathcal{W}(J)$. (For a discussion of the closed ideal structure of $\mathcal{B}(J)$, see [Laus]; Laustsen shows that $\mathcal{K}(J)$ is equal to various other naturally defined ideals.) This gives us an example $\mathfrak{B} = \mathcal{K}(J)$ such that

$$\kappa_{\mathfrak{B}}(\mathfrak{B}) \subsetneq \left(\mathfrak{Z}_t^{(1)}(\mathfrak{B}'') \cap \mathfrak{Z}_t^{(2)}(\mathfrak{B}'')\right). \tag{6.11}$$

We now determine the radicals of $(\mathfrak{A}'', \Box)$ and $(\mathfrak{A}'', \Diamond)$.

Certainly the Banach algebra $(\mathfrak{A}'', \Box) = (\mathcal{B}(E''), \circ)$ is semisimple.

To calculate the radical of $(\mathfrak{A}'', \Diamond)$, we apply the earlier algebraic calculation, with $(\mathfrak{A}'', \circ)$, $\mathcal{B}(E')^a$, $\mathcal{Q}$, and $\mathfrak{J}$ playing the roles of A, B, P, and J, respectively. The product $\Diamond$ in $\mathfrak{A}''$ corresponds to the previously defined product $\times$ in A. Further, $\mathcal{B}(E')^a$ is a semisimple subalgebra of $(\mathfrak{A}'', \circ)$, and

$$\mathfrak{J} = \{U \in \mathcal{B}(E'') : U \mid E = 0\}$$

is a left ideal in $(\mathfrak{A}'', \circ)$, and so conditions (1), (2), and (3) of those algebraic conditions are satisfied. We conclude that

$$\operatorname{rad}(\mathfrak{A}'', \diamond) = \ker \mathcal{Q} \quad \text{and} \quad (\mathfrak{A}'', \diamond) = \mathcal{B}(E')^a \ltimes \operatorname{rad}(\mathfrak{A}'', \diamond),$$

so that $(\mathfrak{A}'', \diamond)$ is strongly decomposable. In particular, the radicals of $(\mathfrak{A}'', \Box)$ and $(\mathfrak{A}'', \diamond)$ are not the same. □

EXAMPLE 6.3. Let $\mathfrak{A} = \mathcal{K}(c_0)$, as above. Then we have

$$\mathfrak{Z}_t^{(1)}(\mathfrak{A}'') \neq \mathfrak{Z}_t^{(2)}(\mathfrak{A}'') \quad \text{and} \quad \operatorname{rad}(\mathfrak{A}'', \Box) \neq \operatorname{rad}(\mathfrak{A}'', \diamond).$$

For $x = (x_n) \in c_0$, set $\overline{x} = (\overline{x}_n)$, and then, for $T \in \mathfrak{A}$, set

$$\overline{T}x = \overline{T\overline{x}} \quad (x \in \mathfrak{A}).$$

Then $\overline{T} \in \mathfrak{A}$, and the map $T \mapsto \overline{T}$ is a linear involution on $\mathfrak{A}$ such that $\overline{ST} = \overline{S}\,\overline{T}$ $(S, T \in \mathfrak{A})$.

As in Example 4.4, we can construct a Banach $*$-algebra $\mathfrak{C} = \mathfrak{A} \oplus \mathfrak{A}^{\mathrm{op}}$ such that $\mathfrak{Z}_t^{(1)}(\mathfrak{C}'') \neq \mathfrak{Z}_t^{(2)}(\mathfrak{C}'')$ and $\operatorname{rad}(\mathfrak{C}'', \Box) \neq \operatorname{rad}(\mathfrak{C}'', \diamond)$. □

[*Added in August, 2004:* In his thesis at Leeds [Da2], Daws has now substantially extended the above results; he deals with more general Banach spaces and obtains more definitive results and further examples.]

CHAPTER 7

Beurling Algebras

In this chapter, we shall describe the Banach algebras that we shall consider. First, we recall some standard notation; see [D] for further information.

Let S be a set, and let $s \in S$. We write both δ_s and λ_s for the characteristic function of the singleton $\{s\}$. Let $\omega : S \to \mathbb{R}^{+\bullet}$ be a function. Then:

$$\ell^1(S,\omega) = \left\{ f = \sum_{s \in S} f(s)\delta_s : \|f\|_\omega = \sum_{s \in S} |f(s)|\,\omega(s) < \infty \right\};$$

$$\ell^\infty(S, 1/\omega) = \left\{ \lambda = \sum_{s \in S} \lambda(s)\lambda_s : \|\lambda\|_{\infty,\omega} = \sup_{s \in S} \frac{|\lambda(s)|}{\omega(s)} < \infty \right\}.$$

Clearly $(\ell^1(S,\omega), \|\cdot\|_\omega)$ and $\left(\ell^\infty(S,1/\omega), \|\cdot\|_{\infty,\omega}\right)$ are Banach spaces, and the latter is the dual of the former for the pairing

$$(f, \lambda) \mapsto \langle f,\, \lambda \rangle = \sum_{s \in S} f(s)\lambda(s)\,.$$

Note that $\omega \in \ell^\infty(S, 1/\omega)$. We also define

$$c_0(S, 1/\omega) = \{\lambda \in \ell^\infty(S, 1/\omega) : |\lambda|\,/\omega \in c_0(S)\}\,.$$

Then $c_0(S, 1/\omega)$ is a closed subspace of $\ell^\infty(S, 1/\omega)$ containing λ_s for each $s \in G$ and

$$c_0(S, 1/\omega) = \overline{\mathrm{lin}}\,\{\lambda_s : s \in S\};$$

the dual space of $c_0(S, 1/\omega)$ is $\ell^1(S,\omega)$, and the second dual space is identified with $\ell^\infty(S, 1/\omega)$. In the case where $\omega = 1$, we write $\ell^1(S)$, etc. Note that $\|\delta_s\|_\omega = \omega(s)$ and $\|\lambda_s\|_{\infty,\omega} = 1/\omega(s)$ for $s \in G$.

Let G be a locally compact group. We denote by m a fixed left Haar measure on G, and by Δ_G the modular function of G. The group G with the discrete topology is denoted by G_d. Let G and H be locally compact groups. Then the product group $G \times H$ is also a locally compact group. For details on locally compact groups, see [HR1] and [RS].

Let X be a Banach space of measures or of equivalence classes of functions on a locally compact group G, and then let $\omega : G \to \mathbb{R}^{+\bullet}$ be a continuous function. We define

$$X(\omega) = \{f : \omega f \in X\}.$$

The norm of $X(\omega)$ is defined so that the map $f \mapsto \omega f$ from $X(\omega)$ onto X is a linear isometry. In particular, we define

$$L^1(G,\omega) = \left\{ f \text{ Borel measurable} : \|f\|_\omega = \int_G |f(s)|\,\omega(s)\,\mathrm{d}m(s) < \infty \right\}$$

and

$$L^\infty(G, 1/\omega) = \left\{ \lambda \text{ Borel measurable} : \|\lambda\|_{\infty,\omega} = \operatorname*{ess\,sup}_{s\in G} \frac{|\lambda(s)|}{\omega(s)} < \infty \right\}.$$

We identify two functions f and g in $L^1(G,\omega)$ if they are equal almost everywhere with respect to m, and we identify λ and μ in $L^\infty(G, 1/w)$ if they are equal locally almost everywhere with respect to m. Then $(L^1(G,\omega), \|\cdot\|_\omega)$ and $(L^\infty(G, 1/\omega),\ \|\cdot\|_{\infty,\omega})$ are Banach spaces, and the latter is the dual of the former for the pairing

$$(f, \lambda) \mapsto \langle f,\, \lambda\rangle = \int_G f(s)\lambda(s)\,\mathrm{d}m(s)\,;$$

this duality specifies the weak-$*$ topology on $L^\infty(G, 1/\omega)$. We also define:

$$C_0(G, 1/\omega) = \{\lambda \in L^\infty(G, 1/\omega) :\ \lambda/\omega \in C_0(G)\},$$

so that $C_0(G, 1/\omega)$ is a closed subspace of $L^\infty(G, 1/\omega)$. We write $L^1(G)$ and $L^\infty(G)$ in the case where $\omega = 1$; we also write $L^1_{\mathbb{R}}(G,\omega)$ and $L^\infty_{\mathbb{R}}(G, 1/\omega)$, etc., for the real-linear subspaces of $L^1(G,\omega)$ and $L^\infty(G, 1/\omega)$, respectively, consisting of elements which are identified with functions that are real-valued. Finally, we denote by $L^\infty_{00}(G, 1/\omega)$ the subspace of $L^\infty(G, 1/\omega)$ consisting of elements λ such that there is a compact set $K \subset G$ with $\operatorname{supp}\lambda \subset K$.

We shall also utilize the following specific subset of $L^1(G,\omega)$. Define

$$P_\omega(G) = \{f \in L^1(G,\omega) : f \geq 0,\ \|f\|_\omega = 1\}\,;$$

in the case where $\omega = 1$, we write $P(G)$, as in the standard sources (for example, see [Pat, (0.1)]).

We shall use the following classical theorem of Steinhaus.

THEOREM 7.1. *Let $\omega : G \to \mathbb{R}^{+\bullet}$ be a continuous function on a locally compact group G. Then $L^1(G,\omega)$ is a weakly sequentially complete Banach space.* □

Let G be a locally compact group, and let $\omega : G \to \mathbb{R}^{+\bullet}$ be a continuous function. Then the space $L^\infty(G, 1/\omega)$ is a commutative, unital C^*-algebra for the product $\cdot_\omega$ which is defined for $\lambda, \mu \in L^\infty(G, 1/\omega)$ by

$$(\lambda \cdot_\omega \mu)(s) = \lambda(s)\mu(s)/\omega(s) \quad (s \in G);$$

the identity of this algebra is the function ω and the involution is the map $f \mapsto \overline{f}$. Indeed, $L^\infty(G, 1/\omega)$ is a commutative von Neumann algebra. Clearly $L^\infty(G, 1/\omega)$ is $*$-isomorphic to the C^*-algebra $L^\infty(G)$, and $C_0(G, 1/\omega)$ is a C^*-subalgebra of $L^\infty(G, 1/\omega)$.

An element $\lambda \in L^\infty(G, 1/\omega)$ is self-adjoint (respectively, positive) as an element of the C^*-algebra if and only if λ is identified with a function that takes its values in $\mathbb{R}$ (respectively, in $\mathbb{R}^+$).

In the case where $\omega(s) \geq 1 \ (s \in G)$, we denote by $M(G, \omega)$ the Banach space of all complex-valued, regular Borel measures μ on G such that

$$\|\mu\|_\omega = \int_G \omega(s) \, \mathrm{d} |\mu| (s) < \infty,$$

and we write $M(G)$ in the case where $\omega = 1$, so that $M(G, \omega)$ is a subspace of $M(G)$ and $M(G, \omega)$ is the dual of $C_0(G, 1/\omega)$ for the pairing

$$(\lambda, \mu) \mapsto \langle \lambda, \mu \rangle = \int_G \lambda(s) \, \mathrm{d}\mu(s);$$

this latter duality defines the weak-$*$ topology on $M(G, \omega)$.

There is a decomposition of $M(G, \omega)$. Let $M_a(G, \omega)$ and $M_s(G, \omega)$ denote the closed linear subspaces of $M(G, \omega)$ consisting of those measures which are absolutely continuous and singular, respectively, with respect to the Haar measure. Then $M(G, \omega) = M_a(G, \omega) \oplus M_s(G, \omega)$ as an ℓ^1-sum of Banach spaces. We identify $M_a(G, \omega)$ with the closed subspace $L^1(G, \omega)$ of $M(G, \omega)$; the space $M_s(G, \omega)$ contains the closed subspace $\ell^1(G, \omega)$ of discrete measures in $M(G, \omega)$. (Here we regard each element δ_s as a measure on G by setting $\delta_s(E) = 1$ whenever $s \in E$ and $\delta_s(E) = 0$ whenever $s \notin E$ for a Borel set E in G.) This theory is essentially that given in [HR1] in the case where $\omega = 1$. We shall just use the Banach space decomposition

$$M(G, \omega) = L^1(G, \omega) \oplus M_s(G, \omega). \tag{7.1}$$

The following result is standard.

PROPOSITION 7.2. *The closed subspaces $L^1(G, \omega)$ and $\ell^1(G, \omega)$ are weak-$*$ dense in $M(G, \omega)$.*

PROOF. Set $A = L^1(G,\omega)$ and $M = M(G,\omega)$, so that

$$A'' = L^\infty(G, 1/\omega)'.$$

Take $\mu \in M$ with $\|\mu\| \leq 1$. By the Hahn–Banach theorem, there exists $\overline{\mu} \in A''$ with $\|\overline{\mu}\| \leq 1$ and $\overline{\mu} \mid C_0(G,1/\omega) = \mu$. There is a net (f_α) in $A_{[1]}$ such that $f_\alpha \to \overline{\mu}$ in the topology $\sigma(A'', A')$. Clearly $f_\alpha \to \mu$ in $(M, \sigma(M, C_0))$. This gives the result for the subspace $L^1(G,\omega)$.

A similar argument gives the result for the subspace $\ell^1(G,\omega)$. □

Let G be a locally compact group. We shall now recall the definitions of some standard closed subspaces of the space $(L^\infty(G), \|\cdot\|_\infty)$; in the case where we are dealing with continuous functions on G, we denote the norm $\|\cdot\|_\infty$ by $|\cdot|_G$, so that $|\cdot|_G$ is the uniform norm. In particular, each algebra $(\ell^\infty(G), \cdot)$ is a commutative, unital von Neumann algebra, identified with $C(\beta G_d)$, and the Banach space $\ell^1(G)''$ is identified with $M(\beta G_d)$.

We shall utilize the *left* and *right translations* ℓ_t and r_t, defined for functions f on G and $t \in G$ by the formulae:

$$(\ell_t f)(s) = f(ts), \quad (r_t f)(s) = f(st) \quad (s \in G). \tag{7.2}$$

(The functions $\ell_t f$ and $r_t f$ are denoted by ${}_t f$ and f_t in [HR1, Chapter 15]; the left shift $S_t f$ is defined by $(S_t f)(s) = f(t^{-1}s)$ $(s \in G)$ in [D].)

For $\lambda \in \mathbb{C}^G$, we set

$$LO(\lambda) = \{\ell_t\lambda : t \in G\}, \quad RO(\lambda) = \{r_t\lambda : t \in G\}.$$

Thus:

$CB(G)$ denotes the closed subspace of $L^\infty(G)$ consisting of the (equivalence classes of) bounded, continuous functions on G (as in Chapter 2);

$LUC(G)$ and $RUC(G)$ denote the closed subspaces of $CB(G)$ consisting of the (equivalence classes of) *bounded, left* (respectively, *right*) *uniformly continuous functions* on G, so that

$$\begin{aligned} LUC(G) &= \{\lambda \in CB(G) : t \mapsto \ell_t\lambda,\ G \to CB(G),\ \text{is continuous}\} \\ RUC(G) &= \{\lambda \in CB(G) : t \mapsto r_t\lambda,\ G \to CB(G),\ \text{is continuous}\}; \end{aligned}$$

$WAP(G)$ denotes the closed subspace of $CB(G)$ consisting of the *weakly almost periodic functions*, so that

$$WAP(G) = \{\lambda \in CB(G) : LO(\lambda) \text{ is relatively compact in the weak topology of } CB(G)\};$$

$AP(G)$ denotes the closed subspace of $CB(G)$ consisting of the *almost periodic functions*, so that

$$AP(G) = \{\lambda \in CB(G) \ : \ LO(\lambda) \text{ is relatively compact in the norm topology of } CB(G)\}\,.$$

Note that elements of $LUC(G)$ are called 'right uniformly continuous' in [HR1]. In fact,

$$\begin{aligned} LUC(G) &= \{\lambda \in L^\infty(G) : t \mapsto \ell_t\lambda\,,\ G \to L^\infty(G),\ \text{is continuous}\} \\ RUC(G) &= \{\lambda \in L^\infty(G) : t \mapsto r_t\lambda\,,\ G \to L^\infty(G),\ \text{is continuous}\}\,. \end{aligned}$$

This follows from Proposition 7.15, below.

It is well-known (see [BJM, pp. 130, 139] and [HR1, Theorem (18.1)]) that $\lambda \in WAP(G)$ (respectively, $\lambda \in AP(G)$) if and only if $RO(\lambda)$ is relatively compact in the weak (respectively, norm) topology of $CB(G)$. Further, it is known (see [BJM, pp. 128, 138]) that

$$(7.3)\quad AP(G) = AP(G_d) \cap C(G)\,, \quad WAP(G) = WAP(G_d) \cap C(G)\,.$$

Let $\lambda \in L^\infty(G)$. Then

$$(7.4)\qquad \overline{\langle\{\ell_t\lambda : t \in G\}\rangle} = \overline{\{\lambda \cdot f : f \in P(G)\}}\,,$$

where the closures are taken in the weak-$*$ topology (see [Wo, Lemma 6.3]). In particular, in the case where $\lambda \in WAP(G)$, the weak and weak-$*$ topologies coincide on the $\|\cdot\|$-closure of $\langle\{\ell_t\lambda : t \in G\}\rangle$, and hence the sets in (7.4) are $\|\cdot\|$-closed, and so must be contained in $LUC(G)$. It follows easily from (7.4) that

$$WAP(G) = \{\lambda \in CB(G) : \lambda \cdot P(G) \text{ is relatively weakly compact}\}\,,$$

and so

$$(7.5)\qquad AP(G) = AP(L^1(G))\,, \quad WAP(G) = WAP(L^1(G))\,.$$

Further, it follows that

$$\begin{aligned} WAP(G) &= \{\lambda \in L^\infty(G) \ : \ LO(\lambda) \text{ is relatively compact in the weak topology of } L^\infty(G)\}\,, \\ AP(G) &= \{\lambda \in L^\infty(G) \ : \ LO(\lambda) \text{ is relatively compact in the norm topology of } L^\infty(G)\}\,, \end{aligned}$$

and that

$$L^\infty(G) \supset CB(G) \supset LUC(G) \supset WAP(G) \supset AP(G)\,.$$

We also have $WAP(G) \supset C_0(G)$; however, in the case where G is locally compact and non-compact, $AP(G) \cap C_0(G) = \{0\}$ [BJM, Chapter 4, Corollary 1.15].

The following result is immediate from Theorem 3.3; see also [BJM, Chapter 4, Theorem 2.3].

THEOREM 7.3. *Let G be a locally compact group, and take an element $\lambda \in CB(G)$. Then $\lambda \in WAP(G)$ if and only if the function $(s,t) \mapsto \lambda(st)$, $G \times G \to \mathbb{C}$, clusters on $G \times G$.* □

It was proved by Granirer [Gra2, p. 62] that $LUC(G) = WAP(G)$ if and only if the group G is compact; see also [L3, Corollary 4].

DEFINITION 7.4. *Let $\omega : G \to \mathbb{R}^{+\bullet}$ be a continuous function. Then:*

$$\begin{aligned} CB(G,1/\omega) &= \{\lambda \in L^\infty(G,1/\omega) : \lambda/\omega \in CB(G)\}\,; \\ LUC(G,1/\omega) &= \{\lambda \in L^\infty(G,1/\omega) : \lambda/\omega \in LUC(G)\}\,; \\ RUC(G,1/\omega) &= \{\lambda \in L^\infty(G,1/\omega) : \lambda/\omega \in RUC(G)\}\,; \\ WAP(G,1/\omega) &= \{\lambda \in L^\infty(G,1/\omega) : \lambda/\omega \in WAP(G)\}\,; \\ AP(G,1/\omega) &= \{\lambda \in L^\infty(G,1/\omega) : \lambda/\omega \in AP(G)\}\,. \end{aligned}$$

We clearly have

$$\begin{aligned} L^\infty(G,1/\omega) &\supset CB(G,1/\omega) \supset LUC(G,1/\omega) \\ &\supset WAP(G,1/\omega) \supset AP(G,1/\omega)\,. \end{aligned}$$

Also $CB(G,1/\omega) \subset \ell^\infty(G,1/\omega)$ and $C_0(G,1/\omega) \subset WAP(G,1/\omega)$. In the case where G is discrete, we have

$$CB(G,1/\omega) = LUC(G,1/\omega) = \ell^\infty(G,1/\omega)\,,$$

and, in the case where G is compact, we have

$$CB(G,1/\omega) = AP(G,1/\omega)\,;$$

further, $AP(G,1/\omega) = WAP(G,1/\omega)$ only if G is compact.

The spaces

$$CB(G,1/\omega), \quad WAP(G,1/\omega), \quad AP(G,1/\omega), \quad LUC(G,1/\omega)$$

are each a C^*-subalgebra of $L^\infty(G,1/\omega)$ and each space contains the function ω.

We now introduce a central concept of this memoir, that of a weight on a group.

DEFINITION 7.5. *Let G be a group, with identity e_G. A* weight *on G is a function $\omega : G \to \mathbb{R}^{+\bullet}$ such that*

$$\omega(st) \le \omega(s)\omega(t) \quad (s,t \in G), \quad \omega(e_G) = 1\,.$$

A function ω on G is symmetric *if*

$$\omega(s^{-1}) = \omega(s) \quad (s \in G)\,.$$

Let $\omega : G \to \mathbb{R}^{+\bullet}$ be a function such that

$$\omega(st) \leq \omega(s)\omega(t) \quad (s, t \in G)\,.$$

Then $\omega(e_G) \geq 1$. By changing the value of ω at e_G to be 1, we obtain a weight on G. Notice that

$$\omega(s^{-1})^{-1} \leq \omega(st)/\omega(t) \leq \omega(s) \quad (s, t \in G)\,. \tag{7.6}$$

Let G be a group, and let $\omega : G \to \mathbb{R}^{+\bullet}$ be a function. It is often convenient to set

$$\omega = \exp \eta\,,$$

so that $\eta : G \to \mathbb{R}$ is a function, and ω is a weight if and only if η is subadditive and $\eta(e_G) = 0$. Of course, $\omega(s) \geq 1\ (s \in G)$ if and only if $\eta(s) \geq 0\ (s \in G)$. Further, for a function $\eta : G \to \mathbb{R}$, we shall set

$$(\delta^1\eta)(s, t) = \eta(s) - \eta(st) + \eta(t) \quad (s, t \in G)\,. \tag{7.7}$$

We also adopt throughout the following notation. Following [CrY], we write

$$\Omega(s, t) = \frac{w(st)}{w(s)w(t)} \quad (s, t \in G)\,, \tag{7.8}$$

so that $0 < \Omega(s, t) \leq 1\ (s, t \in G)$ and $\Omega = \exp(-\delta^1\eta)$ as functions on $G \times G$.

For example, let G be a group, and let S and T be disjoint subsets of G such that $S \cup T = G \setminus \{e_G\}$. Define

$$\eta(e_G) = 0\,, \quad \eta(s) = 2 \ \ (s \in S)\,, \quad \eta(t) = 1 \ \ (t \in T)\,. \tag{7.9}$$

Then η satisfies the above conditions, and $\omega = \exp \eta$ is a weight on G.

THEOREM 7.6. *Let ω be a weight on a group G. Then the Banach space $\ell^1(G, \omega)$ is a unital Banach algebra with respect to the convolution product $\star$, defined by the requirement that*

$$\delta_s \star \delta_t = \delta_{st} \quad (s, t \in G)\,.$$

□

DEFINITION 7.7. *Let ω be a weight on a group G. The algebras $\ell^1(G, \omega)$ are the (discrete)* Beurling algebras *on G.*

The dual module of $\ell^1(G, \omega)$ is identified with $\ell^\infty(G, 1/\omega)$. Note that, in this case, we have

$$\lambda_s \cdot \delta_t = \lambda_{t^{-1}s}, \quad \delta_t \cdot \lambda_s = \lambda_{st^{-1}} \quad (s, t \in G)\,, \tag{7.10}$$

where we are regarding λ_s as an element of $\ell^\infty(G, 1/\omega)$. Let $t \in G$. Then

$$\ell_t\lambda = \lambda \cdot \delta_t, \quad r_t\lambda = \delta_t \cdot \lambda \quad (\lambda \in \ell^\infty(G, 1/\omega))\,.$$

For example, take $G = (\mathbb{Z}, +)$, and write $\ell^1(\omega)$ for $\ell^1(\mathbb{Z}, \omega)$. For each $\alpha \geq 0$, define

$$\omega_\alpha(n) = (1 + |n|)^\alpha \quad (n \in \mathbb{Z})\,. \tag{7.11}$$

Then each ω_α is a weight on $\mathbb{Z}$, and $\ell^1(\omega_\alpha)$ is a Beurling algebra on $\mathbb{Z}$. It is easy to see that the corresponding function Ω_α 0-clusters strongly on $G \times G$ whenever $\alpha > 0$.

Let G be an abelian group with (compact) dual group Γ, and let ω be a weight on G with $\omega(s) \geq 1$ $(s \in G)$. The Fourier transform identifies $\ell^1(G, \omega)$ as a Banach function algebra on Γ, and so $\ell^1(G, \omega)$ is a semisimple Banach algebra. In particular, $\ell^1(\omega_\alpha)$ is identified with a subalgebra of $C(\mathbb{T})$; this latter algebra is contained in the Banach function algebra $\mathrm{lip}_\alpha \mathbb{T}$. In the case where $\alpha \geq 1$, each such Fourier transform is continuously differentiable on $\mathbb{T}$. For details of these remarks, see [D, Example 4.6.13].

For a compact subset X of $\mathbb{C}$, denote by $A(X)$ the uniform algebra of all continuous functions on X that are analytic on the interior of X. Let ω be a weight on the group $(\mathbb{Z}, +)$ with $\omega(n) \geq 1$ $(n \in \mathbb{Z})$, and set

$$\rho_1 = \inf\{\omega_n^{1/n} : n \in \mathbb{N}\}, \quad \rho_2 = \sup\{\omega_{-n}^{-1/n} : n \in \mathbb{N}\}\,,$$

so that $0 < \rho_2 \leq 1 \leq \rho_1 < \infty$. Then the character space of $\ell^1(\omega)$ is homeomorphic to the annulus

$$X := \{z \in \mathbb{C} : \rho_2 \leq |z| \leq \rho_1\}\,,$$

and $\widehat{f} \in A(X)$ for each $f \in \ell^1(\omega)$ [D, p. 504].

DEFINITION 7.8. *Let G be a locally compact group. A* weight function *on G is a continuous function $\omega : G \to \mathbb{R}^{+\bullet}$ such that ω is a weight on G.*

For example, let η be a continous, subadditive function on G such that $\eta(e_G) = 0$. Then $\omega = \exp \eta$ is a weight function on G.

Let ω_1 and ω_2 be weight functions on a locally compact group G. Then the pointwise product $\omega_1\omega_2$ is also a weight function on G. Let ω_1 and ω_2 be weight functions on locally compact groups G_1 and G_2, respectively. Define $\omega_1 \otimes \omega_2$ on $G_1 \times G_2$ by

$$(\omega_1 \otimes \omega_2)(s, t) = \omega_1(s)\omega_2(t) \quad (s \in G_1,\, t \in G_2)\,.$$

Then $\omega_1 \otimes \omega_2$ is a weight function on $G_1 \times G_2$.

Let ω be a weight function on G. Then it follows from (7.6) that

$$\lim_{s \to e_G} \sup_{t \in G} \omega(st)/\omega(t) = 1\,. \tag{7.12}$$

Let $\omega : G \to \mathbb{R}^+$ be a weight function on G. Then the corresponding functions $\delta^1\eta$ and Ω are continuous functions on $G \times G$, and Ω is bounded by 1.

Let ω be a weight function on G with $\omega(s) \geq 1 \ (s \in G)$. Then convolution product $\star$ on $M(G, \omega)$ is defined by the formula

$$\langle \lambda, \mu \star \nu \rangle = \int_G \int_G \lambda(st)\, \mathrm{d}\mu(s)\, \mathrm{d}\nu(t) \quad (\mu, \nu \in M(G, \omega),\ \lambda \in C_0(G, 1/\omega)).$$

THEOREM 7.9. *Let ω be a weight function on a locally compact group G. Then the Banach space $M(G, \omega)$ is a unital Banach algebra with respect to the convolution product $\star$; $L^1(G, \omega)$ is a closed ideal in $M(G, \omega)$, and $\ell^1(G, \omega)$ is a closed subalgebra of $M(G, \omega)$.* □

The product of f and g in $L^1(G, \omega)$ is given by

$$(f \star g)(t) = \int_G f(s)g(s^{-1}t)\, \mathrm{d}m(s) = \int_G f(ts^{-1})g(s)\Delta_G(s^{-1})\, \mathrm{d}m(s)$$

for $t \in G$. We also note the following formulae for the product of $f \in L^1(G)$ and $\mu \in M(G)$:

$$\left.\begin{aligned} (f \star \mu)(t) &= \int_G f(ts^{-1})\Delta_G(s^{-1})\, \mathrm{d}\mu(s)\,, \\ (\mu \star f)(t) &= \int_G f(s^{-1}t)\, \mathrm{d}\mu(s)\,, \end{aligned}\right\} \quad (t \in G)\,,$$

In particular $f \star \delta_s = \Delta_G(s^{-1})r_{s^{-1}}f$ and $\delta_s \star f = \ell_{s^{-1}}f$ for $s \in G$ and $f \in L^1(G)$.

DEFINITION 7.10. *Let ω be a weight function on a locally compact group G. The algebras $L^1(G, \omega)$ are the (continuous)* Beurling algebras *on G.*

For a general background and history on Beurling algebras, see the texts of Reiter and Stegeman [RS] and Palmer [Pa2, 1.9.15].

For example, take $G = (\mathbb{R}, +)$, and define

$$\omega_\alpha(t) = (1 + |t|)^\alpha \quad (t \in \mathbb{R})$$

for $\alpha \geq 0$. Then each ω_α is a weight function on $\mathbb{R}$, and $L^1(\mathbb{R}, \omega_\alpha)$ is a Beurling algebra. Let Ω_α be the corresponding function on $\mathbb{R} \times \mathbb{R}$. Then, in the case where $\alpha > 0$, the function $s \mapsto \Omega_\alpha(s, t)$ is decreasing on $\mathbb{R}^+$ for each $t \in \mathbb{R}$.

We remark in passing that it follows from Theorem 7.1 that 'an infinite-dimensional C^*-algebra cannot be a closed linear subspace of any Beurling algebra'. Indeed, let A be a C^*-algebra, and let B be a Banach algebra which is weakly sequentially complete as a Banach space (for example, let B be a Beurling algebra). Suppose that A is linearly homeomorphic to a closed subspace of B. Then A is also weakly sequentially complete. However the only C^*-algebras which are weakly sequentially complete are those which are finite-dimensional.

Let G be a locally compact group, and let H be the group G with the opposite product, so that H is a locally compact group. For a function f on G, set

$$\check{f}(s) = f(s^{-1}) \quad (s \in G).$$

Now let m be a left Haar measure on G, and define $\check{m}$ by

$$\int f(s)\,\mathrm{d}\check{m}(s) = \int \check{f}(s)\Delta_G(s^{-1})\,\mathrm{dm}(s) \quad (f \in L^1(G)).$$

Then $\check{m}$ is a left Haar measure on H. Let ω be a weight function on G. Then $\check{\omega}/\Delta_G$ is a weight function on H, and the opposite algebra to $L^1(G,\omega)$ is isometrically isomorphic with $L^1(H,\check{\omega}/\Delta_G)$. Thus the opposite algebra to a Beurling algebra is also a Beurling algebra.

DEFINITION 7.11. *Let G be a locally compact group. Two weight functions ω_1 and ω_2 on G are* equivalent *if there exists a continuous algebra isomorphism from $L^1(G,\omega_1)$ onto $L^1(G,\omega_2)$.*

For example, ω_1 and ω_2 are equivalent if there exist constants $c_1 > 0$ and $c_2 > 0$ such that

$$\omega_1(s) \leq c_2\omega_2(s), \quad \omega_2(s) \leq c_1\omega_1(s) \quad (s \in G).$$

Now let ω_1 be a weight and set $\omega_2(n) = \mathrm{e}^{cn}\omega_1(n)$ $(n \in \mathbb{Z})$ for some constant c. Then ω_1 and ω_2 are equivalent. All properties involving topological centres and radicals of second duals are unchanged if we move to an equivalent weight on G.

The Banach algebra $L^1(G,\omega)$ is commutative if and only if G is an abelian group; it has an identity if and only if G is discrete. Suppose that $\omega(s) \geq 1$ $(s \in G)$. Then the algebra $L^1(G,\omega)$ is a dense subalgebra of the group algebra $L^1(G)$.

Let ω be a weight function on a locally compact group G. Then the operators ℓ_t and r_t (for $t \in G$) act on $L^\infty(G,1/\omega)$, and

$$\|\ell_t\| = \|r_t\| = \omega(t) \quad (t \in G).$$

The following remark is clear; we write 1 for the function on G which is constantly equal to 1.

PROPOSITION 7.12. *Let G be locally compact group, and let ω be a weight function on G with $\omega(s) \geq 1$ $(s \in G)$. Then $1 \in L^\infty(G, 1/\omega)$, and the map*

$$\varphi_G : \Phi \mapsto \langle \Phi, 1 \rangle, \quad L^1(G,\omega)'' \to \mathbb{C},$$

is a character on both $(L^1(G,\omega)'', \Box)$ and $(L^1(G,\omega)'', \Diamond)$. $\Box$

The above character φ_G on $(L^1(G,\omega)'', \Box)$, $(L^1(G,\omega)'', \Diamond)$, and their subalgebras is the *augmentation character*; its kernel is the *augmentation ideal.*

Again let G be a locally compact group. Then there is an isometric involution $*$ on $M(G)$, defined by the formula

$$\mu^*(E) = \overline{\mu(E^{-1})} \quad (\mu \in M(G))$$

for each Borel subset E of G. Thus we have $\delta_s^* = \delta_{s^{-1}}$ $(s \in G)$ and

$$f^*(s) = \overline{f(s^{-1})}\Delta_G(s^{-1}) \quad (f \in L^1(G),\, s \in G).$$

The algebra $(M(G), *)$ is a Banach $*$-algebra. For $\mu \in M(G)$, define

$$T_\mu : f \mapsto \mu \star f, \quad H \to H,$$

where H is the Hilbert space $L^2(G)$ and

$$(\mu \star f)(t) = \int_G f(s^{-1}t)\, \mathrm{d}\mu(s) \quad (f \in H,\, \mu \in M(G)).$$

Then $T_\mu \in \mathcal{B}(H)$, and the map

$$\mu \mapsto T_\mu, \quad M(G) \to \mathcal{B}(H),$$

is a continuous $*$-isomorphism. It follows that $M(G)$ is $*$-semisimple, and hence semisimple. For this standard theory, see [HR1] and [D, Theorem 3.3.34], etc.

Let G be a locally compact group, and let ω be a weight function on G. We are embarrassed to say that we do not know whether or not the Banach algebra $L^1(G,\omega)$ is always semisimple; in particular, we do not know whether or not $\ell^1(G,\omega)$ is always semisimple. We make some remarks on this question.

It is proved in [BhDe] that $L^1(G,\omega)$ is semisimple in the case where G is abelian (and ω is only required to be measurable).

Let us suppose that G is an arbitrary locally compact group and that ω is a weight function on G such that $\omega(s) \geq 1$ $(s \in G)$, so that $L^1(G,\omega)$ is a subalgebra of $L^1(G)$. (By Theorem 7.44, below, the latter hypothesis is no constraint in the case where G is amenable.)

First, suppose that ω is symmetric. Then $M(G,\omega)$ and $L^1(G,\omega)$ are $*$-subalgebras of $M(G)$ and $L^1(G)$, respectively, (and the involution is an isometry on $M(G,\omega)$), and so both $M(G,\omega)$ and $L^1(G,\omega)$ are $*$-semisimple, and hence semisimple. For a recent study of the Banach $*$-algebras $L^1(G,\omega)$ in the case where ω is symmetric, see [FG].

Second, consider the case where ω is not necessarily symmetric. A group G is said to be *maximally almost periodic* if the continuous, finite-dimensional, irreducible, unitary representations separate the points of G; we also say that '$G \in$ [MAP]'. It is shown in [Pa2, 3.2.17] and [Pa3, 12.4.15] that $G \in$ [MAP] if and only if the algebra $AP(G)$ separates the points of G. The class [MAP] is discussed in [HR1, (22.22)] and [Pa3, Chapter 12.5]; the class includes all discrete free groups, but it does not include the discrete group $SL(n,\mathbb{R})$ in the case where $n \geq 2$. Let $G \in$ [MAP]. Then a short proof given in [Bar] shows that the algebra $M(G)$ has the property that, for each $\mu \in M(G) \setminus \{0\}$, there exists $n \in \mathbb{N}$ and a continuous epimorphism $\theta : M(G) \to \mathbb{M}_n(\mathbb{C})$ such that $\theta(\mu) \neq 0$. (The specific result that we use was also proved earlier in [GM]; see also [Pa2, 12.5.20(b)].) Since each matrix algebra $\mathbb{M}_n(\mathbb{C})$ is simple, we have $\theta(\operatorname{rad} M(G,\omega)) = \{0\}$, and so $\mu \notin \operatorname{rad} M(G,\omega)$. Hence $\operatorname{rad} L^1(G,\omega) = \operatorname{rad} M(G,\omega) = \{0\}$.

Thus we obtain the following result.

THEOREM 7.13. *Let G be a locally compact group, and let ω be a weight function on G such that $\omega(s) \geq 1$ $(s \in G)$. Suppose* either *that G is a maximally almost periodic group* or *that G is arbitrary and ω is a symmetric function. Then $M(G,\omega)$ and $L^1(G,\omega)$ are semisimple.*□

However we cannot prove that the Banach algebra $L^1(G,\omega)$ is semisimple in the case where $G \notin$ [MAP] and ω is not symmetric.

It is standard that $L^1(G)$ always has a bounded approximate identity which is a net consisting of continuous functions of compact support, and this net is clearly also a bounded approximate identity for each of the Beurling algebras $L^1(G,\omega)$. Thus $L^1(G,\omega)$ satisfies (2.2) and $L^1(G,\omega)''$ has a mixed identity, say it is Φ_0; we may suppose that $\|\Phi_0\| = 1$, so that the natural embedding of $L^1(G,\omega)$ into $\mathcal{M}(L^1(G,\omega))$ is an isometry. In the case where G is metrizable, $L^1(G,\omega)$ has a sequential bounded approximate identity.

The next result is a minor extension of a standard theorem of Wendel [We].

THEOREM 7.14. *Let G be a locally compact group, and let ω be a weight function on G with $\omega(s) \geq 1$ $(s \in G)$. Then $\mathcal{M}(L^1(G,\omega))$ is isometrically isomorphic to $M(G,\omega)$. Each multiplier on $L^1(G,\omega)$ has the form (L_μ, R_μ) for some $\mu \in M(G,\omega)$.* □

Here, $L_\mu(f) = \mu \star f$ and $R_\mu(f) = f \star \mu$ for $f \in L^1(G,\omega)$.

It follows that $L^1(G,\omega)$ is a Banach $M(G,\omega)$-bimodule, and so the dual space $L^\infty(G,1/\omega)$ and the second dual space $L^1(G,\omega)''$ are both also Banach $M(G,\omega)$-bimodules. For example, we see that, for each $\lambda \in L^\infty(G,1/\omega)$, we have

$$\lambda \cdot \delta_t = \ell_t\lambda, \quad \delta_t \cdot \lambda = r_t\lambda \quad (t \in G).$$

We next recall the explicit formulae for some module products. Let $f \in L^1(G,\omega)$, $\lambda \in L^\infty(G,1/\omega)$, $\mu \in M(G,\omega)$, and $\Phi \in L^1(G,\omega)''$. Then

$$\langle f,\, \mu \cdot \lambda\rangle = \langle f \star \mu,\, \lambda\rangle, \quad \langle \Phi \cdot \mu,\, \lambda\rangle = \langle \Phi,\, \mu \cdot \lambda\rangle . \tag{7.13}$$

Further, $\mu \cdot \lambda$ and $\lambda \cdot \mu$ can be identified with functions on G by the formulae:

$$\left\{\begin{array}{lcl} (\mu \cdot \lambda)(t) &=& \displaystyle\int_G \lambda(ts)\,\mathrm{d}\mu(s)\,, \\ (\lambda \cdot \mu)(t) &=& \displaystyle\int_G \lambda(st)\,\mathrm{d}\mu(s)\,, \end{array}\right. \tag{7.14}$$

which hold for locally almost all $t \in G$, as in [D, Chapter 3.3] and [HR1, Theorem (20.12)].

The space $C_0(G,1/\omega)$ is regarded as a subspace of $M(G,\omega)'$ through the canonical embedding of $C_0(G,1/\omega)$ in $C_0(G,1/\omega)''$. It is clear that $C_0(G,1/\omega)$ is a closed $M(G,\omega)$-submodule of $M(G,\omega)'$, and so $M(G,\omega)$ is also a dual Banach algebra in the sense of Definition 2.6. However, $L^1(G,\omega)$ is not a dual Banach algebra–indeed, it is not a dual Banach space–unless G is discrete.

Let $\lambda \in L^\infty(G,1/\omega)$. It is clear from (7.6) that $\lambda \in LUC(G,1/\omega)$ if and only if

$$\lim_{s\to e_G} \sup_{t\in G} \frac{|\lambda(st)-\lambda(t)|}{\omega(t)} = 0\,.$$

We shall require the following result of Grønbæk [Gr3, Proposition 1.3]. In fact, there seems to be a gap in the proof given in [Gr3], and so we indicate an argument for this result.

PROPOSITION 7.15. *Let ω be a weight function on a locally compact group G, and let $\lambda \in L^\infty(G,1/\omega)$. Then $\lambda \in LUC(G,1/\omega)$ if and only if the map*

$$s \mapsto \lambda \cdot \delta_s, \quad G \to L^\infty(G,1/\omega)\,, \tag{7.15}$$

is continuous.

PROOF. The point to be clarified is that $\lambda \in CB(G,1/\omega)$ whenever λ satisfies (7.15).

To see this, let $(e_\alpha : \alpha \in A)$ be an approximate identity in $L^1(G,\omega)$ with $\|e_\alpha\| = 1$ for each $\alpha \in A$. Then $\{\lambda \star \check{e}_\alpha : \alpha \in A\}$ is a family of bounded functions on G; we *claim* that it is an equicontinuous family. Indeed, take $s, t \in G$. Then

$$\begin{aligned} |(\lambda \star \check{e}_\alpha)(st) - (\lambda \star \check{e}_\alpha)(t)| &\leq \int_G |\lambda(stu) - \lambda(tu)|\,|e_\alpha(u)|\,\mathrm{d}m(u) \\ &\leq \|\lambda \cdot \delta_{st} - \lambda \cdot \delta_t\|_{\infty,\omega}\,, \end{aligned}$$

and so the claim follows. It is a consequence of Ascoli's theorem that there exists $\mu \in CB(G, 1/\omega)$ with the property that we may suppose that $\lambda \star \check{e}_\alpha \to \mu$ uniformly on compact subsets of G. This implies that $\lambda \star \check{e}_\alpha \to \mu$ in the weak-$*$ topology on $L^\infty(G, 1/\omega)$. But

$$\langle f, \lambda \star \check{e}_\alpha \rangle = \langle f \star e_\alpha, \lambda \rangle \to \langle f, \lambda \rangle \quad (f \in L^1(G, \omega))\,,$$

and so $\lambda \star \check{e}_\alpha \to \lambda$ in the weak-$*$ topology on $L^\infty(G, 1/\omega)$. Hence $\lambda = \mu$ locally almost everywhere (with respect to m) on G. Thus we may indeed suppose that $\lambda \in CB(G, 1/\omega)$.

The remainder of the proof is as in [Gr3, Proposition 1.3]. $\square$

It is important to note that $L^\infty(G, 1/\omega)$ is not an essential $L^1(G,\omega)$-bimodule (unless G be discrete).

We shall now use the following abbreviated notation, which leaves the locally compact group G to be defined implicitly.

Let ω be a weight function on a locally compact group G; from now on, we suppose that $\omega(s) \geq 1$ $(s \in G)$. Then we set

$$\text{(7.16)}\quad \begin{cases} \mathcal{A}_\omega = L^1(G,\omega), \quad \mathcal{A}'_\omega = L^\infty(G, 1/\omega), \quad \mathcal{B}_\omega = \mathcal{A}''_\omega\,, \\ \mathcal{M}_\omega = M(G,\omega), \ \ \mathcal{X}_\omega = LUC(G, 1/\omega), \ \ \mathcal{E}_\omega = C_0(G, 1/\omega)\,, \\ \mathcal{W}_\omega = WAP(G, 1/\omega), \quad \mathcal{AP}_\omega = AP(G, 1/\omega)\,, \end{cases}$$

so that $\mathcal{E}'_\omega = \mathcal{M}_\omega$ as a Banach space. We shall also write $\mathcal{S}_\omega$ for the state space of the C^*-algebra $\mathcal{A}'_\omega$. Of course, we always take $\mathcal{M}_\omega$ and $\mathcal{A}_\omega$ to be Banach algebras for the product $\star$; $\mathcal{B}_\omega$ is a Banach algebra for the two products $\Box$ and $\Diamond$.

PROPOSITION 7.16. *Let ω be a weight function on a locally compact group G. Then the spaces $CB(G, 1/\omega)$, $\mathcal{X}_\omega$, $\mathcal{E}_\omega$, $\mathcal{W}_\omega$, and $\mathcal{AP}_\omega$ are C^*-subalgebras of $\mathcal{A}'_\omega$, and each space is translation-invariant and an $\mathcal{A}_\omega$-bimodule. The modules $CB(G, 1/\omega)$, $\mathcal{X}_\omega$, $\mathcal{E}_\omega$, and $\mathcal{W}_\omega$ are faithful.*

PROOF. This result is immediate. In fact, each A-bimodule X of $\mathcal{A}'_\omega$ such that $\mathcal{E}_\omega \subset X$ is faithful. $\square$

Note that it is not the case that $\mathcal{W}_\omega = WAP(L^1(G,\omega))$ for an arbitrary weight function ω on a locally compact group G, as we shall see below in equation (9.1).

PROPOSITION 7.17. *Let ω be a weight function on a locally compact group G. Then:*

(i) $\mathcal{A}'_\omega \cdot \mathcal{A}_\omega = \mathcal{X}_\omega \cdot \mathcal{A}_\omega = \mathcal{X}_\omega$;
(ii) $\mathcal{A}_\omega \cdot \mathcal{A}'_\omega = RUC(G, 1/\omega)$;
(iii) $\mathcal{A}_\omega \cdot \mathcal{E}_\omega = \mathcal{E}_\omega \cdot \mathcal{A}_\omega = \mathcal{E}_\omega$;
(iv) $L^\infty_{00}(G, 1/\omega) \cdot \mathcal{A}_\omega \subset \mathcal{E}_\omega$.

PROOF. (i), (ii), and (iii) These are standard because $\mathcal{A}_\omega$ has a bounded approximate identity (*cf.* [HR1, (20.19)], [HR2, Chapter 32], and [Gr3, Proposition 1.3]).

(iv) Let $f \in \mathcal{A}_\omega$, and let $\lambda \in \mathcal{A}'_\omega$ be such that $\operatorname{supp}\lambda \subset K$ for a compact subset K of G. Fix $\varepsilon > 0$, and choose a compact subset L of G such that $\int_{G\setminus L} |f(u)|\,\mathrm{d}m(u) < \varepsilon$. Then

$$|(f \cdot \lambda)(t)| < \varepsilon \,\|\lambda\|\,\omega(t) \quad (t \in G \setminus (L^{-1} \cdot K)),$$

and so $f \cdot \lambda \in \mathcal{E}_\omega$. □

Clause (iii) of the above result says that $\mathcal{E}_\omega$ is a neo-unital Banach $\mathcal{A}_\omega$-bimodule.

Let A be a Banach algebra with a bounded approximate identity. We noted in Proposition 3.12 that $WAP(A) \subset A' \cdot A$. Consider the special case where A is $L^1(G)$ for a locally compact group G. Then $A' \cdot A$ is equal to $LUC(G)$ and $WAP(A) = WAP(G)$, and we have remarked that it is a theorem of Granirer that $WAP(G) = LUC(G)$ only if G is compact. Thus, again, we have examples of Banach algebras A with $WAP(A) \neq A' \cdot A$.

PROPOSITION 7.18. *Let ω be a weight function on a locally compact group G. Then $\mathcal{W}_\omega$ and $\mathcal{AP}_\omega$ are neo-unital Banach $\mathcal{A}_\omega$-bimodules.*

PROOF. Let $\mathcal{W}$ and $\mathcal{AP}$ denote $\mathcal{W}_\omega$ and $\mathcal{AP}_\omega$, respectively, in the special case where $\omega = 1$. It follows from (7.5) and Proposition 3.12 that $\mathcal{W}$ and $\mathcal{AP}$ are neo-unital Banach $\mathcal{A}_\omega$-bimodules.

Let $(e_\alpha) \subset P(G)$ be an approximate identity for $L^1(G)$; we may suppose that supp e_α is eventually contained in each compact neighbourhood of e_G, and so (e_α) is a bounded appproximate identity for $\mathcal{A}_\omega$.

Let $\lambda \in \mathcal{W}_\omega$ with $\|\lambda\|_{\infty,\omega} \leq 1$, say. Then $\lambda/\omega \in \mathcal{W}$, and so we have $(\lambda/\omega) \cdot e_\alpha \to \lambda/\omega$ uniformly on G. We *claim* that $\lambda \cdot e_\alpha \to \lambda$ in $\mathcal{W}_\omega$;

for this, we must show that $(\lambda \cdot e_\alpha)/\omega \to \lambda/\omega$ uniformly on G. and so it suffices to show that

$$\lim_\alpha |(\lambda/\omega) \cdot e_\alpha - (\lambda \cdot e_\alpha)/\omega|_G = 0\,. \tag{7.17}$$

Fix $\varepsilon > 0$. By (7.12), there is a compact neighbourhood K of e_G such that $|1 - \omega(st)/\omega(t)| < \varepsilon$ for each $t \in G$ and $s \in K$. For each α such that $\operatorname{supp} e_\alpha \subset K$, we have

$$\begin{aligned} &|(\lambda/\omega) \cdot e_\alpha - (\lambda \cdot e_\alpha)/\omega|_G \\ &\quad = \sup_{t \in G} \left| \int_K \left(\frac{\lambda(st)}{\omega(st)} - \frac{\lambda(st)}{\omega(t)} \right) e_\alpha(s)\, \mathrm{d}m(s) \right| \\ &\quad \le \sup_{t \in G} \int_K |1 - \omega(st)/\omega(t)|\, e_\alpha(s)\, \mathrm{d}m(s) < \varepsilon\,, \end{aligned}$$

and so the result follows.

This calculation shows that $\mathcal{W}_\omega = \overline{\mathcal{W}_\omega \cdot \mathcal{A}_\omega}$; similarly, we have $\mathcal{W}_\omega = \overline{\mathcal{A}_\omega \cdot \mathcal{W}_\omega}$, and so $\mathcal{W}_\omega$ is an essential Banach $\mathcal{A}_\omega$-bimodule. By Theorem 2.3(ii), $\mathcal{W}_\omega$ is neo-unital.

Similarly, $\mathcal{AP}_\omega$ is neo-unital. □

THEOREM 7.19. *Let ω be a weight function on a locally compact group G. Then:*

(i) *$\mathcal{X}_\omega$ is left-introverted as a subspace of both $\mathcal{A}'_\omega$ and $\ell^\infty(G, 1/\omega)$;*

(ii) *$\mathcal{E}_\omega$ is introverted as a subspace of both $\mathcal{A}'_\omega$ and $\ell^\infty(G, 1/\omega)$.*

PROOF. (i) First, $\mathcal{X}_\omega$ is left-introverted in $\mathcal{A}'_\omega$ because $\mathcal{X}_\omega = \mathcal{A}'_\omega \cdot \mathcal{A}_\omega$ by Proposition 7.17(i).

We show directly that $\mathcal{X}_\omega$ is a left-introverted subspace of A'_ω, where $A'_\omega = \ell^\infty(G, 1/\omega)$. Take $\lambda \in \mathcal{X}_\omega \subset A'_\omega$, and $\Phi \in A''_\omega$, so that $\Phi \cdot \lambda \in A'_\omega$. Let $s_\alpha \to s$ in G. Then

$$\|(\Phi \cdot \lambda) \cdot \delta_{s_\alpha} - (\Phi \cdot \lambda) \cdot \delta_s\| \le \|\Phi\|\, \|\lambda \cdot \delta_{s_\alpha} - \lambda \cdot \delta_s\| \to 0\,,$$

and so $\Phi \cdot \lambda \in \mathcal{X}_\omega$ by Proposition 7.15.

(ii) We regard $\mathcal{E}_\omega$ as a subspace of either $\mathcal{A}'_\omega$ or $\ell^\infty(G, 1/\omega)$. Fix $\lambda \in \mathcal{E}_\omega$ and $\mu \in \mathcal{E}'_\omega = \mathcal{M}_\omega$. Then the formulae for $\mu \cdot \lambda$ and $\lambda \cdot \mu$ are given in (7.14); it is clear that $\mu \cdot \lambda$ and $\lambda \cdot \mu$ have compact support whenever both μ and λ have compact support, and so $\mu \cdot \lambda, \lambda \cdot \mu \in \mathcal{E}_\omega$ in this special case. The general case follows. Thus, by Proposition 5.2, $\mathcal{E}_\omega$ is introverted in both $\mathcal{A}'_\omega$ and $\ell^\infty(G, 1/\omega)$. □

Part of the above proposition follows from our general result, Theorem 5.15, applied with $A = \mathcal{A}_\omega$, with $B = \ell^1(G, \omega)$, and with $X = \mathcal{X}_\omega$. By Theorem 7.14, the multiplier algebra $\mathcal{M}(A)$ is $\mathcal{M}_\omega = M(G, \omega)$;

clearly, B is an $\|\cdot\|$-closed, unital subalgebra of $\mathcal{M}(A)$ and also B is $\sigma(X', X)$-dense in X'. In this case, X is a faithful, left-introverted submodule of B'. Theorem 5.15 gives us one further piece of information that we isolated as a proposition; the result follows from equation (5.6).

PROPOSITION 7.20. *Let ω be a weight function on a locally compact group G. Then the product $\Box$ in $\mathcal{X}'_\omega$ is the same whether $\mathcal{X}_\omega$ be regarded as a subspace of $\mathcal{A}'_\omega$ or of $\ell^\infty(G, 1/\omega)$.* $\Box$

The above result (in the case where $\omega = 1$) was proved by direct calculation in [L1, Lemma 3]. For further information on the Banach algebras $(\mathcal{X}_\omega, \Box)$ in this case, see [GhLaa], [GhL1], [L3], and [LLos1]. The finite-dimensional ideals in this algebra are considered in [Fi1] and [Fi2]; it is interesting that there is a marked difference between finite-dimensional left ideals and finite-dimensional right ideals.

Note also that there is a partial converse to Theorem 7.19 in the case where $\omega = 1$. Let X be a translation-invariant subspace of $CB(G)$ such that X is left-introverted as a subspace of $\ell^\infty(G)$. Then $X \subset LUC(G)$ [Mi2]. This implies that $CB(G)$ is not left-introverted as a subspace of $\ell^\infty(G)$ in the case where G is not discrete. We can see this easily for $G = \mathbb{R}$, for example. Indeed, for each $n \in \mathbb{N}$, choose $f_n \in C(\mathbb{R})$ with $f_n(n) = 1$ and $\operatorname{supp} f_n \subset [n - 1/n, n + 1/n]$, and set $f = \sum_{n=1}^\infty f_n$, so that $f \in CB(G)$. Let Φ be an accummulation point of $\{\delta_n : n \in \mathbb{N}\}$ in $\ell^\infty(G)'$. Then we have $(\Phi \cdot f)(0) = 1$, but $(\Phi \cdot f)(t) = 0$ for each $t \in (0, 1)$, and so $\Phi \cdot f \notin CB(G)$.

Let G be a locally compact group. In the case where $\omega = 1$, it is also the case that $AP(G)$ and $WAP(G)$ are introverted subspaces of $L^\infty(G)$; this follows from (7.5) and Proposition 3.12. It does not seem to be obvious that $AP(G, 1/\omega)$ and $WAP(G, 1/\omega)$ are introverted subspaces of $\mathcal{A}'_\omega$ in the general case; for a partial result on this, see Proposition 11.3.

We make a remark on the C^*-algebras X of the form $AP(G)$, $WAP(G)$, $LUC(G)$, $CB(G)$, and $L^\infty(G)$. Each has a compact character space, say Φ_X, and there are continuous surjections

$$\Phi_{L^\infty(G)} \to \Phi_{CB(G)} \to \Phi_{LUC(G)} \to \Phi_{WAP(G)} \to \Phi_{AP(G)} .$$

Of course, $\Phi_{CB(G)}$ has been identified with βG, the Stone-Čech compactification of G. For a study of the character space of the Banach algebra $L^\infty(G)$, see [LMPy].

Let X be a C^*-subalgebra of $LUC(G)$. Then $\delta_s \in \Phi_X$ $(s \in G)$ and the function $s \mapsto \delta_s \cdot \lambda$ belongs to $LUC(G)$ for each $\lambda \in X$, and so we can make the following calculation. Let $\Psi \in X'$, $\lambda, \mu \in X$, and $s \in G$.

Then we have $\Psi \cdot \lambda \in LUC(G)$ and

$$\begin{aligned}\langle \delta_s, \Psi \cdot (\lambda\mu)\rangle &= \langle \Psi, \delta_s \cdot (\lambda\mu)\rangle = \langle \Psi, (\delta_s \cdot \lambda)(\delta_s \cdot \mu)\rangle \\ &= \langle \Psi, \delta_s \cdot \lambda\rangle\langle \Psi, \delta_s \cdot \mu\rangle = \langle \delta_s, \Psi \cdot \lambda\rangle\langle \delta_s, \Psi \cdot \mu\rangle\,,\end{aligned}$$

and so $\Psi \cdot (\lambda\mu) = (\Psi \cdot \lambda)(\Psi \cdot \mu)$. Now suppose that $\Phi, \Psi \in \Phi_X$, $\lambda, \mu \in X$, and $s \in G$. Then

$$\begin{aligned}\langle \Phi \Box \Psi, \lambda\mu\rangle &= \langle \Phi, \Psi \cdot (\lambda\mu)\rangle = \langle \Phi, (\Psi \cdot \lambda)(\Psi \cdot \mu)\rangle \\ &= \langle \Phi, \Psi \cdot \lambda\rangle\langle \Phi, \Psi \cdot \mu\rangle = \langle \Phi \Box \Psi, \lambda\rangle\langle \Phi \Box \Psi, \mu\rangle\,,\end{aligned}$$

and so $\Phi \Box \Psi \in \Phi_X$. Thus the compact space Φ_X is a semigroup for the product $(\Phi, \Psi) \mapsto \Phi \Box \Psi$. In the case where X contains $C_0(G)$, the map $s \mapsto \delta_s$ is an embedding of G into Φ_X and the range of this map is dense in Φ_X, and so Φ_X is a compactification of G. In the case where $X = WAP(G)$, the product $(\Phi, \Psi) \mapsto \Phi \Box \Psi$ is separately continuous [BJM, Chapter 4, Theorem 2.11]. Next, in the case where $X = LUC(G)$, the product $(\Phi, \Psi) \mapsto \Phi \Box \Psi$ is not necessarily separately continuous. In the case where $X = AP(G)$, the product $(\Phi, \Psi) \mapsto \Phi \Box \Psi$ is jointly continuous and $\Phi_{AP(G)}$ is a compact group [BJM, Chapter 4, Corollary 1.12]; however, it may be that $\Phi_{AP(G)}$ is a singleton. Since $WAP(G) \supset C_0(G)$, it is easy to see that $\Phi_{WAP(G)}$ contains the one-point compactification of G as a homeomorphic image; in the case where G is a non-compact, simple, connected Lie group with a finite centre, $\Phi_{WAP(G)}$ is just this one-point compactification (see [Rup, Theorem 3.6.3]). For a general theory of the compactification of semigroups, see [BJM].

In fact, the spaces $\Phi_{LUC(G)}$, $\Phi_{WAP(G)}$, and $\Phi_{AP(G)}$ are called the *LUC-compactification*, the *WAP-compactification*, and the *AP-compactification* or *Bohr compactification* of G, respectively, and they are the semigroup compactifications of G that are universal with respect to being 'right topological semigroups', 'semitopological semigroups', and 'topological semigroups', respectively. One can define them in this way and prove their existence without using any terminology of C^*-algebras. For this approach, see Chapter 21 of the book [HiSt].

There is a natural definition of the *topological centre* of these semigroups (see [HiSt, Definition 2.4]; the topological centre of $\Phi_{LUC(G)}$ is G itself [LMiPy]. Again let $(S, +)$ be a semigroup. Then βS is a subset of $\ell^1(S)''$ and the two products $\Box$ and $\Diamond$ on $\ell^1(S)''$ give products such that $(\beta S, \Box)$ and $(\beta S, \Diamond)$ are semigroups with operations extending that of G. The *centre* of βS is defined to be

$$\Lambda(S) = \{s \in \beta S : s \Box t = t \Box s \quad (t \in \beta S)\}\,.$$

It is proved in [HiSt, Theorem 6.54] that $\Lambda(S)$ is equal to the centre of S whenever S is 'weakly left cancellative'; in particular, in the case where G is an abelian group, $\Lambda(G) = G$.

We introduce two further sets at this time.

DEFINITION 7.21. *Let ω be a weight function on a locally compact group G. Then*

$$\mathcal{R}_\omega^\Box = \mathrm{rad}\,(\mathcal{B}_\omega,\,\Box\,), \quad \mathcal{R}_\omega^\diamond = \mathrm{rad}\,(\mathcal{B}_\omega,\,\diamond\,)\,.$$

In the case where G is abelian, we see that $\mathcal{R}_\omega^\Box$ and $\mathcal{R}_\omega^\diamond$ are anti-isomorphic as algebras and equal as subsets of $\mathcal{B}_\omega$, and this set is denoted just by $\mathcal{R}_\omega$. In the case where ω is symmetric on G, $\mathcal{R}_\omega^\Box$ and $\mathcal{R}_\omega^\diamond$ are anti-isomorphic as algebras because the extension of the involution $*$ on $\mathcal{A}_\omega$ to $\mathcal{B}_\omega$ maps $\mathcal{R}_\omega^\Box$ onto $\mathcal{R}_\omega^\diamond$. However, we do not know whether or not they are necessarily the same subset of $\mathcal{B}_\omega$. In particular, we do not know whether or not the radicals of the two basic algebras $(\ell^1(G)'',\,\Box\,)$ and $(\ell^1(G)'',\,\diamond\,)$ are the same set for each group G; we guess that this is not the case. Finally, suppose that G is not abelian and that ω is not symmetric: we shall see in Theorem 10.12 that $\mathcal{R}_\omega^\Box$ and $\mathcal{R}_\omega^\diamond$ may be neither isomorphic nor anti-isomorphic as algebras, but we do not know if they are ever distinct subsets of $\mathcal{B}_\omega$.

Thes radicals will be discussed further; for the case where G is discrete, see Chapter 8, and for the case where G is not discrete, see Chapter 12.

THEOREM 7.22. *Let ω be a weight function on a locally compact group G. Then $\mathcal{X}_\omega^\circ$ and $\mathcal{E}_\omega^\circ$ are closed ideals in $(\mathcal{B}_\omega,\,\Box\,)$, and $\mathcal{X}_\omega^\circ$ is a left-annihilator ideal with $\mathcal{X}_\omega^\circ \subset \mathcal{R}_\omega^\Box$.*

PROOF. This follows from Theorem 5.4 because $\mathcal{X}_\omega$ and $\mathcal{E}_\omega$ are left-introverted in $\mathcal{A}_\omega'$ and $\mathcal{X}_\omega = \mathcal{A}_\omega' \cdot \mathcal{A}_\omega$. □

The Banach algebra $\mathcal{X}_\omega'$ is a quotient of $\mathcal{B}_\omega$. Indeed, $\mathcal{X}_\omega' = \mathcal{B}_\omega/\mathcal{X}_\omega^\circ$; we denote the quotient map by q_ω. Clearly $q_\omega(\mathcal{E}_\omega^\circ) = \mathcal{E}_\omega^\circ/\mathcal{X}_\omega^\circ$ is a closed ideal in $(\mathcal{X}_\omega',\,\Box\,)$.

COROLLARY 7.23. *Let ω be a weight function on a non-discrete, locally compact group G, and let Φ_0 be a mixed identity for $\mathcal{B}_\omega$. Then:*

(i) $\mathcal{B}_\omega = \mathcal{X}_\omega' \ltimes \mathcal{X}_\omega^\circ$;

(ii) $\mathcal{X}_\omega' = \Phi_0 \,\Box\, \mathcal{B}_\omega$;

(iii) $q_\omega\left(\mathfrak{Z}_t^{(1)}(\mathcal{B}_\omega)\right) \subset \mathfrak{Z}_t(\mathcal{X}_\omega')$;

(iv) $\mathfrak{Z}_t^{(2)}(\mathcal{B}_\omega) \subset \Phi_0 \,\Box\, \mathcal{B}_\omega$;

(v) $\Phi_0 \notin \mathfrak{Z}_t^{(1)}(\mathcal{B}_\omega)$, *and* $\mathcal{A}_\omega$ *is not Arens regular.*

PROOF. This follows from Proposition 5.9 and Theorem 2.21 once we know that $\mathcal{A}'_\omega \cdot \mathcal{A}_\omega = \mathcal{X}_\omega$. □

COROLLARY 7.24. *Let* ω *be a weight function on a locally compact abelian group* G, *and let* Φ_0 *be a mixed identity for* $\mathcal{B}_\omega$. *Then*

$$\mathcal{B}_\omega = \mathcal{X}'_\omega \ltimes \mathcal{X}^\circ_\omega$$

and

$$q_\omega(\mathfrak{Z}(\mathcal{B}_\omega)) \subset \mathfrak{Z}(\mathcal{X}'_\omega), \quad \mathfrak{Z}(\mathcal{B}_\omega) \subset \Phi_0 \,\Box\, \mathcal{B}_\omega, \quad \Phi_0 \notin \mathfrak{Z}(\mathcal{B}_\omega)\,. \tag{7.18}$$

□

Let ω be a weight function on a locally compact group G. We now show that there is an isometric embedding of the multiplier algebra

$$\mathcal{M}_\omega = (M(G,\omega),\, \star)$$

into the Banach algebra $(\mathcal{X}'_\omega,\, \Box\,)$; the result is well-known in the case where $\omega = 1$.

For $\mu \in \mathcal{M}_\omega$, define $\theta\mu \in \mathcal{X}'_\omega$ by

$$\langle \theta\mu, \lambda\rangle = \int_G \lambda(s)\,\mathrm{d}\mu(s) \quad (\lambda \in \mathcal{X}_\omega)\,. \tag{7.19}$$

Clearly $\theta\mu \in \mathcal{X}'_\omega$ and $\theta\mu \mid \mathcal{E}_\omega = \mu$ for each $\mu \in \mathcal{M}_\omega$. In particular, we have $\langle \theta\delta_s,\, \lambda\rangle = \lambda(s)$ for each $s \in G$ and each $\lambda \in \mathcal{X}_\omega$. The map $\theta : \mathcal{M}_\omega \to \mathcal{X}'_\omega$ is a linear isometry.

In equation (5.5), a map $\theta : \mathcal{M}(A) \to X'$ was defined for a Banach algebra A with an approximate identity of bound 1 and a faithful, left-introverted submodule X of A'. Let us apply this formula in the special case where $A = \mathcal{A}_\omega$, so that $\mathcal{M}(A) = \mathcal{M}_\omega$, and where $X = \mathcal{X}_\omega$, so that $X = X \cdot A$ by Proposition 7.17(i). For each $f \in \mathcal{A}_\omega$, $\lambda \in \mathcal{X}_\omega$, and $\mu \in \mathcal{M}_\omega$, formula (5.5) gives $\langle \theta\mu,\, \lambda \cdot f\rangle = \langle f \star \mu, \lambda\rangle$, whereas the above formula (7.19) gives

$$\begin{aligned}
\langle \theta\mu,\, \lambda \cdot f\rangle &= \int_G (\lambda \cdot f)(s)\,\mathrm{d}\mu(s) \\
&= \int_G \left(\int_G f(t)\lambda(ts)\,\mathrm{d}m(t)\right)\mathrm{d}\mu(s) \\
&= \int_G \left(\int_G f(ts^{-1})\Delta(s^{-1})\,\mathrm{d}\mu(s)\right)\lambda(t)\,\mathrm{d}m(t) \\
&= \langle f \star \mu, \lambda\rangle\,.
\end{aligned}$$

We have shown that the two formulae are consistent. Thus we obtain the following result from Theorem 5.14.

THEOREM 7.25. *The map* $\theta : (\mathcal{M}_\omega, \star) \to (\mathcal{X}'_\omega, \Box)$ *is a continuous embedding, and* $\theta(\mathcal{M}_\omega) \subset \mathfrak{Z}_t(\mathcal{X}'_\omega)$. $\Box$

Let Φ_0 be any mixed identity for $\mathcal{B}_\omega$. Then we see that

$$(7.20)\qquad \langle \Phi_0 \cdot \mu, \lambda \rangle = \int_G \lambda(t)\, \mathrm{d}\mu(t) = \langle \theta\mu, \lambda \rangle \quad (\mu \in \mathcal{M}_\omega, \lambda \in \mathcal{X}_\omega).$$

Thus $\Phi_0 \cdot \mu = \theta\mu$ $(\mu \in \mathcal{M}_\omega)$. We note in particular that this equation holds for *each* weak-$*$ accumulation point Φ_0 of the net (e_α).

We now regard $\mathcal{M}_\omega$ as a $\|\cdot\|$-closed subalgebra of $\mathcal{X}'_\omega$, setting

$$(7.21)\qquad \langle \mu, \lambda \rangle = \int_G \lambda(s)\, \mathrm{d}\mu(s) \quad (\mu \in \mathcal{M}_\omega, \lambda \in \mathcal{X}_\omega);$$

we have $\mathcal{M}_\omega \subset \mathfrak{Z}_t(\mathcal{X}'_\omega)$ and

$$(7.22)\qquad \mathcal{X}'_\omega = \mathcal{M}_\omega \ltimes (\mathcal{E}^\circ_\omega / \mathcal{X}^\circ_\omega)$$

in a canonical way.

PROPOSITION 7.26. *Let* ω *be a weight function on a locally compact group, and suppose that the algebra* $\mathcal{M}_\omega$ *is semisimple. Then*

$$\operatorname{rad} \mathcal{X}'_\omega \subset \mathcal{E}^\circ_\omega / \mathcal{X}^\circ_\omega. \qquad \Box$$

We have a similar embedding of $\mathcal{M}_\omega$ in $(\mathcal{B}_\omega, \Box)$. Since $(\mathcal{A}_\omega, \star)$ is a closed ideal in $(\mathcal{M}_\omega, \star)$ (by an embedding which we call ι_ω), we see that ι''_ω is an embedding of $(\mathcal{B}_\omega, \star)$ as a closed ideal in $(\mathcal{M}''_\omega, \star)$. Indeed it follows from (7.1) that

$$\mathcal{M}''_\omega = \mathcal{B}_\omega \oplus M_s(G, \omega)''$$

as a Banach space.

The general theory of Chapter 2 gives a projection

$$P_\omega : (\mathcal{M}''_\omega, \Box) \to (\mathcal{M}_\omega, \star)$$

which is a continuous epimorphism.

DEFINITION 7.27. *The map* $\Pi_\omega : \mathcal{B}_\omega \to \mathcal{M}_\omega$ *is the restriction map* $P_\omega \mid \mathcal{B}_\omega$.

Thus $\Pi_\omega : (\mathcal{B}_\omega, \Box) \to (\mathcal{M}_\omega, \star)$ is a continuous homomorphism.

As in Chapter 4, we have an isometric embedding

$$(7.23)\qquad \kappa_\omega : \mu \mapsto \Phi_0 \cdot \mu = \Phi_0 \Box \mu, \quad (\mathcal{M}_\omega, \star) \to (\mathcal{B}_\omega, \Box),$$

such that $\kappa_\omega(\delta_0) = \Phi_0$ and $\kappa_\omega(\mathcal{M}_\omega) \subset \Phi_0 \cdot \mathcal{B}_\omega$. It is clear that we have $\Phi_0 \Box f = f$ $(f \in L^1(G, \omega))$, and so $\kappa_\omega \mid L^1(G, \omega)$ is the canonical embedding of $L^1(G,\omega)$ into the second dual space $L^1(G,\omega)''$. Thus, using (7.1), we can write

$$\kappa_\omega(M(G,\omega)) = L^1(G,\omega) \oplus \kappa_\omega(M_s(G,\omega))\,. \tag{7.24}$$

It is clear that $\iota''_\omega \circ \kappa_\omega$ is the canonical embedding of $\mathcal{M}_\omega$ into $\mathcal{M}''_\omega$. We also have $q_\omega \circ \kappa_\omega = \theta$, where θ was defined above. (We note that now κ_ω is not canonically defined because there are many mixed identities in $\mathcal{B}_\omega$.)

A form of the following proposition was given by Lamb in [La, Chapter 2.2].

PROPOSITION 7.28. *The map $\Pi_\omega \circ \kappa_\omega$ is the identity map on $\mathcal{M}_\omega$.*

PROOF. Let $\mu \in \mathcal{M}_\omega$ and $\lambda \in \mathcal{E}_\omega \subset \mathcal{X}_\omega$. Then

$$\langle (\Pi_\omega \circ \kappa_\omega)(\mu),\, \lambda\rangle = \langle \Phi_0 \cdot \mu,\, \lambda\rangle = \langle \mu,\, \lambda \cdot \Phi_0\rangle = \langle \mu,\, \lambda\rangle$$

by (6.8), and so $(\Pi_\omega \circ \kappa_\omega)(\mu) = \mu$. The result follows. □

In particular, the map $\Pi_\omega : \mathcal{B}_\omega \to \mathcal{M}_\omega$ is a surjection such that $\mathcal{E}^\circ_\omega = \ker \Pi_\omega$. We thus have the following analogue of (2.17): there is a short exact sequence

$$\sum : 0 \longrightarrow \mathcal{E}^\circ_\omega \longrightarrow (\mathcal{B}_\omega,\, \Box\,) \xrightarrow{\Pi_\omega} (\mathcal{M}_\omega,\, \star\,) \longrightarrow 0$$

of Banach algebras and continuous homomorphisms. The map κ_ω is a splitting homomorphism for $\sum$. Thus we can also write $(\mathcal{B}_\omega,\, \Box\,)$ as the semidirect product.

$$(\mathcal{B}_\omega,\, \Box\,) = \kappa_\omega(\mathcal{M}_\omega) \ltimes \mathcal{E}^\circ_\omega\,. \tag{7.25}$$

Similarly, we have

$$(\mathcal{M}''_\omega,\, \Box\,) = \kappa_\omega(\mathcal{M}_\omega) \ltimes \ker P_\omega\,, \tag{7.26}$$

where $\ker P_\omega$ is the annihilator of $\mathcal{E}_\omega$ in $\mathcal{M}''_\omega$.

We now consider some important, special elements of $\mathcal{B}_\omega$.

DEFINITION 7.29. *Let ω be a weight function on a locally compact group G, and let $\mathrm{M} \in \mathcal{B}_\omega$. Then:*

(i) M *is* left-s-invariant (*for $s \in G$*) *if* $\langle \mathrm{M}, \ell_s\lambda\rangle = \langle \mathrm{M}, \lambda\rangle \;\; (\lambda \in \mathcal{A}'_\omega)$;

(ii) M *is* left-S-invariant (*for a subsemigroup S of G*) *if* M *is left-s-invariant for each $s \in S$;*

(iii) M *is* left-invariant *if it is left-G-invariant;*

(iv) M *is* topologically left-invariant *if*

$$f \cdot \mathrm{M} = \langle f, 1\rangle \mathrm{M} \quad (f \in \mathcal{A}_\omega)\,.$$

For convenience, we now write $s \cdot \Phi$ and $\Phi \cdot s$ for $\delta_s \cdot \Phi$ and $\Phi \cdot \delta_s$, respectively, whenever $s \in G$ and $\Phi \in \mathcal{B}_\omega$. Since $\ell_s\lambda = \lambda \cdot \delta_s$ for each $s \in G$ and $\lambda \in \mathcal{A}'_\omega$, we see that $\mathrm{M} \in \mathcal{B}_\omega$ is left s-invariant if and only if $s \cdot \mathrm{M} = \mathrm{M}$, and that M is left-invariant if and only if

$$s \cdot \mathrm{M} = \mathrm{M} \quad (s \in G)\,.$$

Similarly, we define analogous 'right-invariant' versions of the above concepts.

DEFINITION 7.30. *Let ω be a weight function on a locally compact group G, and let $\mathrm{M} \in \mathcal{B}_\omega$. Then* M *is* invariant *if it is both left-invariant and right-invariant on G.*

Thus M is invariant on G if and only if

$$s \cdot \mathrm{M} = \mathrm{M} \cdot s = \mathrm{M} \quad (s \in G)\,.$$

In the case where $\omega = 1$, an element $\mathrm{M} \in \mathcal{B}_\omega$ is topologically left-invariant if and only if

$$\langle \mathrm{M},\, \lambda \cdot f\rangle = \langle \mathrm{M},\, \lambda\rangle \quad (f \in P(G),\, \lambda \in L^\infty(G))\,,$$

and so the notion coincides with the standard one, given in [Pat, Definition (0.9)], for example.

Of course, in the case where G is a discrete group, an element of $\mathcal{B}_\omega$ is left-invariant if and only if it is topologically left-invariant.

PROPOSITION 7.31. *Let ω be a weight function on a locally compact group G, and let $\mathrm{M} \in \mathcal{B}_\omega$ be topologically left-invariant. Then*

$$\Phi \,\Box\, \mathrm{M} = \langle \Phi,\, 1\rangle \mathrm{M} \quad (\Phi \in \mathcal{B}_\omega)\,.$$

PROOF. Take $\Phi \in \mathcal{B}_\omega$. Then there is a net (f_α) in $\mathcal{A}_\omega$ such that $\lim_\alpha f_\alpha = \Phi$. Since $\lim_\alpha f_\alpha \,\Box\, \mathrm{M} = \Phi \,\Box\, \mathrm{M}$ and $\lim_\alpha \langle f_\alpha, 1\rangle = \langle \Phi, 1\rangle$, the result follows. $\Box$

PROPOSITION 7.32. *Let ω be a weight function on a non-compact, locally compact group G, and let $\mathrm{M} \in \mathcal{B}_\omega$ be topologically left-invariant. Then $\mathrm{M} \in \mathcal{E}_\omega^\circ$.*

PROOF. Take $\lambda \in C_{00}(G, 1/\omega)$ with $\|\lambda\| = 1$, and set $\alpha = \langle \mathrm{M},\, \lambda\rangle$; we shall show that $\alpha = 0$, which is sufficient for the result. We write $L = \mathrm{supp}\ \lambda$.

Choose $f \in C_{00}(G)$ with $\langle f,\, \lambda\rangle = 1$, say $K = \mathrm{supp}\ f$ and $m = |f|_G$.

Let $n \in \mathbb{N}$. Since G is not compact, there exist $s_1, \dots, s_n \in G$ such that the family $\{K^{-1}s_jL : j \in \mathbb{N}_n\}$ is pairwise disjoint. For each

$j \in \mathbb{N}_n$, set $f_j = f \star \delta_{s_j}$, so that $f_j \in C_{00}(G)$ with $|f_j|_G = m$ and $\operatorname{supp} \lambda \cdot f_j \subset K^{-1}s_jL$. Also, $\langle f_j, 1\rangle = \langle f, \lambda\rangle = 1$, and so

$$\langle \mathrm{M}, \lambda \cdot f_j\rangle = \langle f_j, 1\rangle\langle \mathrm{M}, \lambda\rangle = \alpha\,.$$

Set $g = \sum_{j=1}^n f_j$, so that $\lambda \cdot g = \sum_{j=1}^n \lambda \cdot f_j \in C_{00}(G)$. Since the supports of the functions $\lambda \cdot f_1, \dots, \lambda \cdot f_n$ are pairwise disjoint, we have $|g|_G = m$, and so $\|\lambda \cdot g\| \leq m$. Hence

$$n\,|\alpha| = \left|\sum_{j=1}^n \langle \mathrm{M}, \lambda \cdot f_j\rangle\right| = |\langle \mathrm{M}, \lambda \cdot g\rangle| \leq m\,\|\mathrm{M}\|\;.$$

This holds for each $n \in \mathbb{N}$, and so $\alpha = 0$, as required. □

THEOREM 7.33. *Let ω be a weight function on a locally compact group G, and let $\mathrm{M}_1, \mathrm{M}_2 \in \mathcal{B}_\omega$ be topologically left-invariant, with $\{\mathrm{M}_1, \mathrm{M}_2\}$ linearly independent. Then there exist $\alpha, \beta \in \mathbb{C}$ such that $\alpha\mathrm{M}_1 + \beta\mathrm{M}_2 \in \mathcal{R}_\omega^\Box \setminus \mathcal{X}_\omega^\circ$.*

PROOF. There exist $\alpha, \beta \in \mathbb{C}$ such that $\langle\Lambda, 1\rangle = 0$, but $\Lambda \neq 0$, where we set $\Lambda = \alpha\mathrm{M}_1 + \beta\mathrm{M}_2 \in \mathcal{B}_\omega$. It follows from Proposition 7.31 that

$$\Phi \,\Box\, \Lambda = \langle\Phi, 1\rangle\Lambda \quad (\Phi \in \mathcal{B}_\omega)\,;$$

in particular, $\Lambda \,\Box\, \Lambda = 0$.

Let $\Phi \in \mathcal{B}_\omega$. Then

$$(\Phi \,\Box\, \Lambda)^{\Box 2} = \langle\Phi \,\Box\, \Lambda \,\Box\, \Phi, 1\rangle\Lambda = \langle\Phi, 1\rangle^2\langle\Lambda, 1\rangle\Lambda = 0\,,$$

where we are using Proposition 7.12. Thus $\{\Phi \,\Box\, \Lambda : \Phi \in \mathcal{B}_\omega^\#\}$ is a nilpotent left ideal in $(\mathcal{B}_\omega, \Box)$, and so is contained in $\mathcal{R}_\omega^\Box$. In particular, $\Lambda \in \mathcal{R}_\omega^\Box$.

It remains to show that $\Lambda \notin \mathcal{X}_\omega^\circ$. Indeed, assume towards a contradiction that $\Lambda \in \mathcal{X}_\omega^\circ$. Since $\mathcal{X}_\omega = \mathcal{A}_\omega' \cdot \mathcal{A}_\omega$ by Proposition 7.17(i), we have

$$\langle\Lambda, \lambda \cdot f\rangle = 0 \quad (f \in \mathcal{A}_\omega, \lambda \in \mathcal{A}_\omega')\,,$$

and so $f \cdot \Lambda = 0$ $(f \in \mathcal{A}_\omega)$, whence $\Lambda = 0$, a contradiction. □

DEFINITION 7.34. *Let ω be a weight function on a locally compact group. An element $\mathrm{M} \in \mathcal{B}_\omega$ is a* mean *on $\mathcal{A}_\omega'$ if $\mathrm{M} \geq 0$ and $\langle\mathrm{M}, \omega\rangle = 1$.*

The following result is a trivial variant of standard theorems (see [KR, Theorem 4.3.2], for example).

PROPOSITION 7.35. *Let ω be a weight function on a locally compact group G, and let $\mathrm{M} \in \mathcal{B}_\omega$. Then the following conditions on M are equivalent:*

(a) M *is a mean on* $\mathcal{A}'_\omega$*;*

(b) $\mathrm{M} \in \mathcal{S}_\omega$*;*

(c) *for each* $\lambda \in L^\infty_{\mathbb{R}}(G, 1/\omega)$*, we have*

$$\operatorname*{ess\,inf}_{s\in G} \frac{\lambda(s)}{\omega(s)} \leq \langle \mathrm{M},\, \lambda\rangle \leq \operatorname*{ess\,sup}_{s\in G} \frac{\lambda(s)}{\omega(s)}\,.$$

□

We write $\mathcal{L}_{t,\omega}(G)$ for the space of topologically left-invariant means on $\mathcal{A}'_\omega$; in the special case where $\omega = 1$, we write $\mathcal{L}_t(G)$ for this set, as in [Pat2], for example.

DEFINITION 7.36. *A locally compact group G is* amenable *if there is left-invariant mean on $L^\infty(G)$.*

We note that every abelian group and every compact group is amenable. For a discussion of amenable groups, see [D, Chapter 3.3], [Pa2, Chapter 12.5], [Pat2], [Ru2, Chapter 1], and many other sources.

It is not true that there is a left-invariant mean on $\mathcal{A}'_\omega$ for every weight ω on G, even when G is abelian. For let ω be the weight on $\mathbb{Z}$ defined by

$$\omega(n) = \exp(|n|) \quad (n \in \mathbb{Z})\,, \tag{7.27}$$

and assume towards a contradiction that M is a left-invariant mean on $\ell^\infty(\omega)$. Set $\omega_+ = \omega \mid \mathbb{Z}^+$ and $\omega_- = \omega \mid \mathbb{Z}^-$. Then, each for $k \in \mathbb{N}$, we have

$$|\langle \mathrm{M}, \omega_+\rangle| = |\langle \mathrm{M}, \ell_{-k}\omega_+\rangle| \leq \mathrm{e}^{-k}\,,$$

and so $\langle \mathrm{M}, \omega_+\rangle = 0$. Similarly, $\langle \mathrm{M}, \omega_-\rangle = 0$, and so $\langle \mathrm{M}, \omega\rangle = 0$, the required contradiction.

Suppose that there is a left-invariant mean on $\mathcal{A}'_\omega$, and that the weight ω is symmetric. Then there is an invariant mean on $\mathcal{A}'_\omega$. This is shown by a standard argument given in [D, Proposition 3.3.49], for example. However this result may fail in the case where ω is not symmetric. Even in the case where G is amenable and $\omega = 1$, it is not necesssarily the case that every left-invariant mean is right-invariant. Indeed, let G be a discrete group. Then this holds if and only if G is [FC], i.e., each conjugacy class $\{sts^{-1} : s \in G\}$ is finite for each $t \in G$. For this and more general results, see [Pat1].

In the case where $\omega = 1$, each topologically left-invariant mean is left-invariant; however, as proved in [Pat, Chapter 7], for groups G

which are non-discrete and are such that G_d is amenable, there are left-invariant means that are not topologically left-invariant. For each amenable group G, we have $\mathcal{L}_t(G) \neq \emptyset$; as we shall see, in the case where G is also not compact, $\mathcal{L}_t(G)$ is 'large'. The fact that $\mathcal{L}_t(G) \neq \emptyset$ can also be proved by combining Theorems 8.5.4 and 8.6.9 of [RS].

DEFINITION 7.37. *Let ω be a weight function on a locally compact group G. Then ω is* almost left-invariant *if*

$$\operatorname*{Lim}_{t\to\infty} \sup_{s\in K} \left| \frac{\omega(st)}{\omega(t)} - 1 \right| = 0$$

for each compact subset K of G, and almost invariant *if*

$$\operatorname*{Lim}_{t\to\infty} \sup_{s\in K} \left| \frac{\omega(st)}{\omega(t)} - 1 \right| = \operatorname*{Lim}_{t\to\infty} \sup_{s\in K} \left| \frac{\omega(ts)}{\omega(t)} - 1 \right| = 0$$

for each compact subset K of G,

In Example 10.2, we shall show that there are almost left-invariant weights which are not almost invariant.

Let $\omega = \exp \eta$ be a weight on $\mathbb{Z}$. Then ω is almost invariant if and only if $\lim_{j\to\infty} s(j) = 0$ and $\lim_{j\to\infty} s(-j) = 0$, where

$$s(j) = \eta(j+1) - \eta(j) \quad (j \in \mathbb{Z})$$

defines the *slope* s of η. For example, the weight ω_α on $\mathbb{Z}$ is almost invariant for each $\alpha \geq 0$. However, many of the examples of weights ω on $\mathbb{Z}$ to be given in Chapter 9 are not almost invariant.

THEOREM 7.38. *Let G be a locally compact group, and let ω be an almost left-invariant weight function on G. For $\mathrm{M} \in \mathcal{L}_t(G)$, define $\mathrm{M}_\omega \in \mathcal{B}_\omega$ by*

$$\langle \mathrm{M}_\omega, \lambda \rangle = \langle \mathrm{M}, \lambda/\omega \rangle \quad (\lambda \in \mathcal{A}'_\omega)\,. \tag{7.28}$$

Then the map $\mathrm{M} \mapsto \mathrm{M}_\omega$ is a bijection from $\mathcal{L}_t(G)$ onto $\mathcal{L}_{t,\omega}(G)$.

PROOF. Define M_ω as in (7.28). It is immediate from Proposition 7.35 that M_ω is a mean on $\mathcal{A}'_\omega$.

We now show that M_ω is topologically left-invariant. Let $f \in \mathcal{A}_\omega$ and $\lambda \in \mathcal{A}'_\omega$, and set

$$\mu = (\lambda \cdot f)/\omega - (\lambda/\omega) \cdot f\,.$$

Suppose that f has compact support K. For each $t \in G$, we have

$$\mu(t) = \int_K f(s) \cdot \frac{\lambda(st)}{\omega(st)} \left(1 - \frac{\omega(st)}{\omega(t)}\right) \mathrm{d}m(s)\,,$$

and so

$$|\mu(t)| \leq \|f\| \, \|\lambda\| \sup_{s \in K} \left| \frac{\omega(st)}{\omega(t)} - 1 \right| .$$

Since ω is almost left-invariant, we see that $\mu \in C_0(G)$. By Proposition 7.32 (applied in the case where $\omega = 1$), we have $\langle \mathrm{M}, \mu \rangle = 0$. Hence

$$\langle \mathrm{M}_\omega, \lambda \cdot f \rangle = \langle f, 1 \rangle \langle \mathrm{M}_\omega, \lambda \rangle .$$

Since each function $f \in \mathcal{A}_\omega$ is the limit of functions of compact support, it follows that $\langle \mathrm{M}_\omega, \lambda \cdot f \rangle = \langle f, 1 \rangle \langle \mathrm{M}_\omega, \lambda \rangle$ $(f \in \mathcal{A}_\omega,\ \lambda \in \mathcal{A}'_\omega)$, and so $\mathrm{M}_\omega \in \mathcal{L}_{t,\omega}(G)$.

For $\mathrm{M}_\omega \in \mathcal{L}_{t,\omega}(G)$, define M by

$$\langle \mathrm{M}, \lambda \rangle = \langle \mathrm{M}_\omega, \lambda\omega \rangle \quad (\lambda \in L^\infty(G)) .$$

It follows in a similar way to the above that $\mathrm{M} \in \mathcal{L}_t(G)$: now we must show that $(\lambda \cdot f)\,\omega - (\lambda\omega) \cdot f \in \mathcal{E}_\omega$ whenever $f \in L^1(G)$ and $\lambda \in L^\infty(G)$, and this also follows from Proposition 7.32.

The theorem is proved. □

DEFINITION 7.39. *Let Ω be a locally compact space. Then $\kappa(\Omega)$ is the minimal cardinality κ such that there is a family $\{K_i : i \in I\}$ of compact sets with $|I| = \kappa$ such that $\Omega = \bigcup\{K_i : i \in I\}$.*

Clearly $\kappa(\Omega) = 1$ for a compact space Ω, $\kappa(\Omega) = |\Omega|$ for an infinite discrete space Ω, and $\kappa(\Omega) = \aleph_0$ for a σ-compact, non-compact space Ω.

THEOREM 7.40. *Let G be a non-compact, amenable locally compact group, and let ω be an almost left-invariant weight function on G. Then*

$$\dim \mathcal{R}_\omega^\square \geq 2^{2^{\kappa(G)}}, \quad \dim \operatorname{rad} \mathcal{X}'_\omega \geq 2^{2^{\kappa(G)}} .$$

PROOF. Set m$= 2^{2^{\kappa(G)}}$. By a theorem of Lau and Paterson [LPat] (see also [Pat, Theorem (7.6)] and [LMiPy]) we have $|\mathcal{L}_t(G)| = \mathfrak{m}$. By Theorem 7.38,

$$|\mathcal{L}_t(G)| = |\mathcal{L}_{t,\omega}(G)| ,$$

and so $|\mathcal{L}_{t,\omega}(G)| = \mathfrak{m}$. The required conclusion now follows from Theorem 7.33. □

The above theorem was first proved by Granirer in the special case where G is discrete and $\omega = 1$ [Gra1]. See also [L3, Corollary 6]. It seems to be an open question whether or not $\mathcal{R}_\omega^\square \neq \{0\}$ in the case where the group G is not amenable, even when $\omega = 1$. For a discussion of this point, see [Gra2].

In fact, by using a recent result of Filali and Pym [FiPy, Theorem 5], it can be shown in the same way that

$$\dim I \geq 2^{2^{\kappa(G)}}$$

for each non-zero right ideal I in $\mathcal{X}'_\omega$ or $\mathcal{B}_\omega$ in the case where G is a non-compact, locally compact group and ω is an almost left-invariant weight on G. See also [FiSa] for further results. In a similar vein, an earlier result in [LMiPy] showed that there are $2^{2^{\kappa(G)}}$ left ideals in the semigroup $\Phi_{LUC(G)}$ for each non-compact, locally compact group G.

DEFINITION 7.41. *Let $\omega : G \to \mathbb{R}^+$ be a function on a group G, and let S be a subset on G. Then ω is* diagonally bounded *on S if*

$$\sup\left\{\omega(s)\omega(s^{-1}) : s \in S\right\} < \infty\,. \tag{7.29}$$

A symmetric weight is diagonally bounded on S if and only if it is bounded on S.

Let ω be a weight on a locally compact group G. Recall that the opposite algebra to $L^1(G,\omega)$ is $L^1(H,\check{\omega}/\Delta)$, where H is the opposite group to G. Notice that $\check{\omega}/\Delta$ is diagonally bounded on a subset S of G whenever ω is diagonally bounded on S.

The question of the amenability of the algebras $L^1(G,\omega)$ has been studied by Grønbaek [Gr3]. His result is the following.

THEOREM 7.42. *Let ω be a weight function on a locally compact group G. Then $L^1(G,\omega)$ is amenable as a Banach algebra if and only if G is amenable as a locally compact group and ω is diagonally bounded on G.* □

The weak amenability of $L^1(G,\omega)$ is discussed in [Gr1]. One result on this is the following.

THEOREM 7.43. *Let ω be a weight on $\mathbb{Z}$. Then $\ell^1(\omega)$ is weakly amenable if and only if $\inf\{\omega_n\omega_{-n}/n : n \in \mathbb{N}\} < \infty$.* □

For example, let $\omega_\alpha(n) = (1+|n|)^\alpha$ $(n \in \mathbb{Z})$. Then $\ell^1(\omega_\alpha)$ is weakly amenable if and only if $\alpha \leq 1/2$, a result proved earlier in [BCD].

It is often convenient to consider weight functions ω on G such that $\omega(s) \geq 1$ $(s \in G)$ (and we only defined $M(G,\omega)$ in this setting). We now show that, in the case where G is an amenable group, we may suppose that this extra condition always holds. The proof of the following theorem is due to Michael White, and is essentially contained in [Wh].

THEOREM 7.44. *Let G be an amenable locally compact group, and let ω be a weight function on G. Then:*

(i) *there is a continuous function $\chi : G \to \mathbb{R}^{+\bullet}$ such that*

$$\chi(st) = \chi(s)\chi(t) \quad (s,t \in G) \quad \text{and} \quad \chi(s) \le \omega(s) \quad (s \in G);$$

(ii) *there is a weight function $\widetilde{\omega}$ on G such that $\widetilde{\omega}(s) \ge 1$ $(s \in G)$ and $L^1(G, \widetilde{\omega})$ is isometrically isomorphic to $L^1(G, \omega)$.*

PROOF. (i) Let $\omega = \exp \eta$, as before.

For each $s \in G$, define

$$\mu_s(t) = \eta(st) - \eta(t) \quad (t \in G)\,.$$

Then μ_s is a real-valued, continuous function on G, and

$$-\eta(s^{-1}) \le \mu_s(t) \le \eta(s) \quad (t \in G)\,, \tag{7.30}$$

so that $\mu_s \in CB(G) \subset L^\infty(G)$ and $\mu_{e_G} = 0$. For $s_1, s_2 \in G$, we have

$$\eta(s_1 s_2 t) - \eta(t) = (\eta(s_1 s_2 t) - \eta(s_2 t)) + (\eta(s_2 t) - \eta(t)) \quad (t \in G)\,,$$

and so $\mu_{s_1 s_2}(t) = \mu_{s_1}(s_2 t) + \mu_{s_2}(t)$ $(t \in G)$. Hence

$$\mu_{s_1 s_2} = \ell_{s_2}\mu_{s_1} + \mu_{s_2} \in CB(G)\,. \tag{7.31}$$

Let M be a left-invariant mean on G. Then, from (7.31), we have

$$\langle \mathrm{M},\, \mu_{s_1 s_2}\rangle = \langle \mathrm{M},\, \mu_{s_1}\rangle + \langle \mathrm{M}, \mu_{s_2}\rangle \quad (s_1, s_2 \in G)\,,$$

and also, by (7.30), we have

$$-\eta(s^{-1}) \le \langle \mathrm{M},\, \mu_s\rangle \le \eta(s) \quad (s \in G)\,.$$

Now define

$$\chi(s) = \exp \langle \mathrm{M},\, \mu_s\rangle \quad (s \in G)\,.$$

Then $\chi(st) = \chi(s)\chi(t)$ $(s, t \in G)$ and

$$\omega(s^{-1})^{-1} \le \chi(s) \le \omega(s) \quad (s \in G)\,.$$

Let $s_\alpha \to e_G$ in G. Then $\omega(s_\alpha) \to \omega(e_G) = 1$ because ω is continuous, and so $\chi(s_\alpha) \to 1 = \chi(e_G)$. It follows that χ is continuous on G.

Thus $\chi : G \to \mathbb{R}^{+\bullet}$ has all the required properties.

(ii) Define $\widetilde{\omega}(s) = \omega(s)/\chi(s)$ $(s \in G)$. It is immediate that $\widetilde{\omega}$ is a weight function on G and that $\widetilde{\omega}(s) \ge 1$ $(s \in G)$. For $f \in L^1(G, \omega)$, define

$$\theta(f)(s) = f(s)\chi(s) \quad (s \in G)\,.$$

Then $\theta : L^1(G, \omega) \to L^1(G, \widetilde{\omega})$ is an isometric isomorphism.

This concludes the proof of the theorem. □

In particular, the weights ω and $\widetilde{\omega}$ are equivalent. A modification of Example 10.1, below, shows that there is a weight ω on $\mathbb{F}_2$ such that there is no equivalent weight $\widetilde{\omega}$ on $\mathbb{F}_2$ with $\widetilde{\omega}(s) \geq 1 \ (s \in \mathbb{F}_2)$.

The condition that ω be diagonally bounded on the whole of G was considered further by White in [Wh]; the following remarks will be used later. First suppose that G is an amenable group. Then, in this case, the weight function $\widetilde{\omega}$ of Theorem 7.44(ii) is bounded, and so $L^1(G, \omega)$ is already isomorphic to $L^1(G)$.

The following result will be used in Example 10.1 to show that there is an unbounded weight function ω on the free group $\mathbb{F}_2$ such that ω is diagonally bounded on all of $\mathbb{F}_2$.

Let G be a group, and let $\eta : G \to \mathbb{R}$ be a function such that

$$\left|(\delta^1 \eta)(s,t)\right| \leq M \quad (s,t \in G)$$

for a constant $M \geq 0$. Define

$$\omega(s) = \begin{cases} \exp(\eta(s) + M) & (s \in G \setminus \{e_G\}), \\ 1 & (s = e_G). \end{cases}$$

Then $\omega : G \to \mathbb{R}^{+\bullet}$ is a weight, and ω is diagonally bounded on the whole of G.

Suppose that there exists an isomorphism $\theta : \ell^1(G, \omega) \to \ell^1(G)$, and define

$$\chi : s \mapsto (\varphi_G \circ \theta)(\delta_s), \quad G \to \mathbb{C},$$

where $\varphi_G : \sum_{s \in G} \alpha_s \delta_s \mapsto \sum_{s \in G} \alpha_s$ is the augmentation character on $\ell^1(G)$. Then

$$\chi(st) = \chi(s)\chi(t) \quad (s,t \in G)$$

and

$$|\chi(s)| \leq \mathrm{e}^m \omega(s) = \exp(\eta(s) + m + M) \quad (s \in G \setminus \{e_G\}),$$

where $m = \log \|\theta\| \geq 0$. For each $s \in G$, we have

$$\begin{aligned} |\chi(s)| &= \left|\chi(s^{-1})\right|^{-1} \\ &\geq \exp(-\eta(-m) - M) \geq \exp(-m - 2M + \eta(s)) > 0, \end{aligned}$$

and so we can define $\rho(s) = \log |\chi(s)| \ (s \in G)$; further, we have

$$-m - 2M \leq \rho(s) - \eta(s) \leq m + M \quad (s \in G),$$

and so

$$\begin{cases} \rho(st) = \rho(s) + \rho(t) & (s,t \in G), \\ |(\rho - \eta)(s)| \leq m + 2M & (s \in G). \end{cases} \tag{7.32}$$

This equation will be used later.

CHAPTER 8

The Second Dual of $\ell^1(G,\omega)$

In this chapter, we shall take ω to be a weight on a group G, and we shall study the second dual algebras $(\ell^1(G,\omega)'', \Box)$ and $(\ell^1(G,\omega)'', \Diamond)$ of the Beurling algebra $\ell^1(G,\omega)$. Our aim is to give conditions on the weight ω that determine when $\ell^1(G,\omega)$ is Arens regular and when it is strongly Arens irregular. We shall also consider the radicals of these second dual algebras.

Throughout we fix the following notation; it is a small variation of that given in equations (7.16). We set

$$(8.1)\qquad \begin{cases} A_\omega = \ell^1(G,\omega)\,, & A'_\omega = \ell^\infty(G,1/\omega)\,, \\ E_\omega = c_0(G,1/\omega)\,, & B_\omega = A''_\omega\,. \end{cases}$$

Thus E_ω is a closed submodule of A'_ω, and $A_\omega = E'_\omega$. The natural embedding is denoted by $\kappa_\omega : A_\omega \to B_\omega$. As in (2.18), we have

$$(B_\omega, \Box\,) = A_\omega \ltimes E^\circ_\omega$$

as a Banach algebra; the canonical projection from E'''_ω to E'_ω is

$$(8.2)\qquad \pi_\omega : B_\omega \to A_\omega\,,$$

and $\ker\pi_\omega = E^\circ_\omega$. We also usually write $\|\cdot\|$ for both $\|\cdot\|_\omega$ and $\|\cdot\|_{\infty,\omega}$; we regard δ_s as an element of A_ω and λ_s as an element of A'_ω. Throughout, convergence in B_ω is with respect to the weak-$*$ topology, $\sigma(B_\omega, A'_\omega)$, unless we say otherwise.

We shall use the following notation, which is analogous to that in Definition 7.21.

DEFINITION 8.1. *Let ω be a weight on a group G. Then*

$$R^\Box_\omega = \operatorname{rad}(B_\omega, \Box\,)\,, \quad R^\Diamond_\omega = \operatorname{rad}(B_\omega, \Diamond\,)\,.$$

We formally restate the following result because of its importance to us.

PROPOSITION 8.2. *Let ω be a weight on a group G. Then the algebra $A_\omega = \ell^1(G,\omega)$ is a dual Banach algebra, the map*

$$\pi_\omega : (B_\omega, \Box\,) \to (A_\omega, \star)$$

is a continuous epimorphism, and

$$(B_\omega, \Box) = A_\omega \ltimes E_\omega^\circ$$

as a semidirect product. In the case where A_ω is semisimple, we have $R_\omega^\Box \subset E_\omega^\circ$ and $R_\omega^\Diamond \subset E_\omega^\circ$.

PROOF. The predual space of A_ω is E_ω. $\Box$

We have $\pi_\omega(\Phi)(s) = \langle \Phi, \lambda_s \rangle \ (s \in G,\ \Phi \in B_\omega)$. It follows that

$$\Phi \cdot \lambda_s = \sum_{u \in G} \pi_\omega(\Phi)(u^{-1}s)\lambda_u \quad (s \in G)\,, \tag{8.3}$$

the series being convergent in A'_ω. Indeed, for each $s, t \in G$, we have

$$(\Phi \cdot \lambda_s)(t) = \langle \delta_t, \Phi \cdot \lambda_s \rangle = \langle \Phi, \lambda_s \cdot \delta_t \rangle = \langle \Phi, \lambda_{t^{-1}s} \rangle = \pi_\omega(\Phi)(t^{-1}s)\,,$$

and so (8.3) follows.

The *normalized point mass* at $s \in G$ is defined to be $\delta_s/\omega(s)$; it is denoted by $\widetilde{\delta}_s$. Now let S be a subset of the group G. We denote by E_S the set which is the closure in B_ω of the set

$$\left\{\widetilde{\delta}_s : s \in S\right\}.$$

Clearly, $E_S \cap A_\omega = \left\{\widetilde{\delta}_s : s \in S\right\}$, and $E_S \not\subset A_\omega$ in the case where S is infinite.

DEFINITION 8.3. *Let ω be a weight on a group G, and let S be a subset of G. Then $A_\omega(S) = \ell^1(S, \omega)$, regarded as a closed linear subspace of A_ω, and $B_\omega(S)$ is the weak-$*$ closure of $A_\omega(S)$ in B_ω.*

Thus $A_\omega(S)$ is weak-$*$ dense in B_ω, and $B_\omega(S)$ is the weak-$*$ closed linear span of E_S. In the case where S is a subsemigroup of G, $A_\omega(S)$ and $B_\omega(S)$ are subalgebras of A_ω and B_ω, respectively.

As before, the space A'_ω is a commutative, unital C^*-algebra for the product $\cdot_\omega$; A'_ω is $*$-isomorphic to the C^*-algebra $\ell^\infty(G)$. The state space of A'_ω is denoted by S_ω, so that

$$S_\omega = \{\Phi \in B_\omega : \|\Phi\| = \langle \Phi, \omega \rangle = 1\}\,.$$

For each $s \in G$, define $\varphi_s(\lambda) = \lambda(s)/\omega(s) \ (\lambda \in A'_\omega)$. The function φ_s is a character on A'_ω, and φ_s can be identified with $\widetilde{\delta}_s$. Let Δ_ω be the character space of A'_ω (with the Gel'fand topology), so that Δ_ω is a compact subspace of B_ω. The map $s \mapsto \varphi_s$, $G \to \Delta_\omega$, is an embedding, and clearly $\Delta_\omega = \operatorname{ex} S_\omega$ is homeomorphic to βG; the algebra A'_ω can be identified as a C^*-algebra with $C(\Delta_\omega)$, and B_ω is $M(\Delta_\omega)$ as a Banach space. For each subset S of G, E_S is a subset of Δ_ω. Since $\overline{\langle G \rangle} = S_\omega$, it follows from the converse to the Krein–Milman theorem that $E_G = \Delta_\omega$.

In the case where $\omega = 1$, the set $\left\{\widetilde{\delta}_s : s \in G\right\}$ is a subsemigroup of $(B_\omega, \Box)$, and so Δ_ω is also a subsemigroup $(B_\omega, \Box)$. However, this is not necessarily the case for an arbitrary weight ω.

Let ω be a weight on a group G, and let $f \in A_\omega$. Then we say that $f \geq 0$ if $f(s) \geq 0 \ (s \in G)$, and then

$$A_\omega^+ = \{f \in A_\omega : f \geq 0\} ,$$

so that A_ω^+ is a cone in $(A_\omega, +)$. Note that $f \star g \in A_\omega^+$ whenever $f, g \in A_\omega^+$.

Now take $\lambda \in A'_\omega$. Then we say that $\lambda \geq 0$ if $\langle f, \lambda\rangle \geq 0 \ (f \in A_\omega^+)$, and then

$$(A'_\omega)^+ = \{\lambda \in A'_\omega : \lambda \geq 0\} ,$$

so that $(A'_\omega)^+$ is a cone in $(A'_\omega, +)$. Further, we have $\lambda \geq 0$ if and only if $\lambda(s) \geq 0 \ (s \in G)$. Clearly $(A'_\omega)^+$ is exactly the cone of positive elements in the C^*-algebra A'_ω.

Finally, take $\Phi \in B_\omega$. Then $\Phi \geq 0$ if $\langle \Phi, \lambda\rangle \geq 0 \ (\lambda \in (A'_\omega)^+)$, and then

$$B_\omega^+ = \{\Phi \in B_\omega : \Phi \geq 0\} ,$$

so that B_ω^+ is a cone in $(B_\omega, +)$.

Suppose that $f \in A_\omega$. Then $\kappa_\omega(f) \in B_\omega^+$ if and only if $f \in A_\omega^+$. The cone B_ω^+ is identified with the cone of positive measures in $M(\Delta_\omega)$.

For $\Phi \in B_\omega$, define $\Phi^\triangleleft \in B_\omega$ by setting

$$\langle \Phi^\triangleleft, \lambda\rangle = \overline{\langle \Phi, \overline{\lambda}\rangle} \quad (\lambda \in A'_\omega) ;$$

the element Φ is *hermitian* if $\Phi = \Phi^\triangleleft$ (*cf.* [KR, p. 255]). Let $\Phi \in B_\omega$ be hermitian. Then $\Phi = \Phi^+ - \Phi^-$, where $\Phi^+, \Phi^- \in B_\omega^+$ and

$$\|\Phi\| = \left\|\Phi^+\right\| + \left\|\Phi^-\right\| .$$

The elements Φ^+ and Φ^- are uniquely specified by these conditions. For arbitrary $\Phi \in B_\omega$, define $\Phi^+ = ((\Phi + \Phi^\triangleleft)/2)^+$. Clearly

$$B_\omega^+ = \{\Phi^+ : \Phi \in B_\omega\} .$$

We note that $f^\triangleleft = \overline{f}$ for $f \in A_\omega$.

The following results are standard (and follow easily from the above remarks).

LEMMA 8.4. *Let $\Phi \in B_\omega$. Then there exist $\Phi_1, \dots, \Phi_4 \in B_\omega^+$ such that $\Phi = \Phi_1 - \Phi_2 + \mathrm{i}(\Phi_3 - \Phi_4)$, such that $\Phi_1 = \Phi^+$, and such that $\|\Phi_j\| \leq \|\Phi\| \ (j = 1, \dots, 4)$.* □

The above decomposition is just the Hahn decomposition of measures in $M(\Delta_\omega)$.

LEMMA 8.5. *The subspace B_ω^+ is closed in B_ω, and A_ω^+ is dense in B_ω^+.* □

For example, $\delta_s \in A_\omega^+$ $(s \in G)$, and so $\Delta_\omega \subset B_\omega^+$.

LEMMA 8.6. *Let $\Phi \in B_\omega$ and $\Psi \in B_\omega^+$. Then $(\Phi \,\Box\, \Psi)^+ = \Phi^+ \,\Box\, \Psi$ and $(\Phi \diamond \Psi)^+ = \Phi^+ \diamond \Psi$.*

PROOF. Let $\Phi = \lim_\alpha f_\alpha$, where (f_α) is a net in A_ω. Then clearly $\Phi^+ = \lim_\alpha f_\alpha^+$. Also, let (g_β) be a net in A_ω^+ such that $\Psi = \lim_\beta g_\beta$. We have

$$(f_\alpha \star g_\beta)^+ = f_\alpha^+ \star g_\beta$$

for each α and β. Since $\Phi \,\Box\, \Psi = \lim_\alpha \lim_\beta f_\alpha \star g_\beta$, we have

$$(\Phi \,\Box\, \Psi)^+ = \lim_\alpha \lim_\beta (f_\alpha \star g_\beta)^+ = \lim_\alpha \lim_\beta f_\alpha^+ \star g_\beta = \Phi^+ \,\Box\, \Psi\,.$$

The justification for the formula $(\Phi \diamond \Psi)^+ = \Phi^+ \diamond \Psi$ is similar. □

We remark that it is certainly not true that $(\Phi \,\Box\, \Psi)^+ = \Phi^+ \,\Box\, \Psi^+$ for arbitrary $\Phi, \Psi \in B_\omega$.

We can now give a useful proposition.

PROPOSITION 8.7. *Let ω be a weight on a group G. Assume that, for each $\Phi \in B_\omega^+ \setminus A_\omega$, there exists $\Psi \in \Delta_\omega$ such that $\Phi \,\Box\, \Psi \neq \Phi \diamond \Psi$. Then A_ω is left strongly Arens irregular.*

PROOF. Take $\Phi \in B_\omega \setminus A_\omega$, and let Φ have the decomposition

$$\Phi = \Phi_1 - \Phi_2 + \mathrm{i}(\Phi_3 - \Phi_4)$$

of Lemma 8.4, so that $\Phi_1 = \Phi^+$. By replacing Φ by $c\Phi$ for a suitable $c \in \mathbb{C}$, we may suppose that $\Phi^+ \notin A_\omega$. By hypothesis, there exists $\Psi \in \Delta_\omega$ with $\Phi^+ \,\Box\, \Psi \neq \Phi^+ \diamond \Psi$. But now $(\Phi \,\Box\, \Psi)^+ \neq (\Phi \diamond \Psi)^+$ by Lemma 8.6, and so $\Phi \,\Box\, \Psi \neq \Phi \diamond \Psi$. Thus A_ω is left strongly Arens irregular. □

The seminal study of the Banach algebra $(\ell^1(\mathbb{Z})'', \Box)$ was given by Civin and Yood in [CiY], and their results are a guide to us. However, there is a clear distinction; as proved by Craw and Young [CrY], the Beurling algebra $\ell^1(\omega_\alpha)$ is Arens regular if and only if $\alpha > 0$. We shall prove that $\ell^1(\omega_\alpha)$ is Arens regular in the case where $\alpha > 0$ by a somewhat different method. In fact, the theorem proved by Craw and Young [CrY, Theorem 1] gives necessary and sufficient condition for the Arens regularity of a weighted group algebra; the proof in [CrY] uses some rather general compactness results from [Y1], and we wish to give an elementary proof of a slightly stronger result that is applicable in our specific situation.

THEOREM 8.8. *Let ω be a weight function on a group G, and let S and T be infinite subsets of G. Suppose that Ω 0-clusters on $S \times T$. Then $\Phi \,\Box\, \Psi = 0$ whenever $\Phi \in B_\omega(S) \cap E_\omega^\circ$ and $\Psi \in B_\omega(T) \cap E_\omega^\circ$.*

PROOF. Take $\Phi \in B_\omega(S) \cap E_\omega^\circ$ and $\Psi \in B_\omega(T) \cap E_\omega^\circ$; we may suppose that $\|\Phi\| = \|\Psi\| = 1$. Let $\lambda \in \ell^\infty(G)$. Then, by Proposition 3.1, we may choose sequences (f_m) and (g_n) in $\ell^1(S)$ and $\ell^1(T)$, respectively, such that $\|f_m\| = \|g_n\| = 1$ in each case and such that

$$\lim_m \lim_n \sum_{s\in S,\, t\in T} f_m(s)g_n(t)\lambda(st)\Omega(s,t) = \langle \Phi \,\Box\, \Psi,\, \lambda\omega\rangle . \tag{8.4}$$

Set $K_m = \operatorname{supp} f_m$ $(m \in \mathbb{N})$ and $L_n = \operatorname{supp} g_n$ $(n \in \mathbb{N})$. Then we may suppose that each set K_m and L_n is finite. Also, we have

$$\lim_m f_m(s) = \lim_n g_n(s) = 0 \quad (s \in G)\,,$$

and so we may suppose that

$$K_{n+1} \cap (K_1 \cup \cdots \cup K_n) = L_{n+1} \cap (L_1 \cup \cdots \cup L_n) = \emptyset \quad (n \in \mathbb{N})\,.$$

Set $H = \bigcup_m K_m \cup \bigcup_n L_n$, a countable set. By replacing S and T by $S \cap H$ and $T \cap H$, respectively, we may suppose that S and T are countable.

Set $\Omega_t(s) = \Omega(s,t)$ $(s \in S,\ t \in T)$, and regard $\mathcal{F} := \{\Omega_t : t \in T\}$ as a subset of $C(\beta S)$. For $n \in \mathbb{N}$, define

$$\varphi_n = \sum \{|g_n(t)|\, \Omega_t : t \in L_n\}\,,$$

so that $\varphi_n \in \langle \mathcal{F} \rangle$. Since Ω clusters on $S \times T$, it follows from Theorem 3.3 that $\langle \mathcal{F} \rangle$ is relatively weakly sequentially compact, and so we may suppose, on passing to a subsequence of (φ_n), that $\varphi_n \to h \in C(\beta S)$.

Fix $x \in \beta S \setminus S$. It follows from Proposition 3.5 that, for each $\varepsilon > 0$, we have $|\varphi_n(x)| < \varepsilon$ for all but finitely many $n \in \mathbb{N}$, and so $|h(x)| \le \varepsilon$. This proves that $h(x) = 0$, and so $h \mid (\beta S \setminus S) = 0$.

Again fix $\varepsilon > 0$. Then there exists $m_0 \in \mathbb{N}$ with

$$|h|_{K_m} < \varepsilon \quad (m \ge m_0)\,,$$

and so

$$\begin{aligned} &\lim_m \lim_n \sum \{|f_m(s)g_n(t)\lambda(st)|\, \Omega(s,t) : s \in K_m,\ t \in L_n\} \\ &\qquad = \lim_m \sum_{s\in K_m} |f_m(s)h(t)|\, \|\lambda\|_\infty \le \varepsilon \|\lambda\|_\infty\,. \end{aligned}$$

This holds for each $\varepsilon > 0$, and so $\langle \Phi \,\Box\, \Psi,\, \lambda\omega\rangle = 0$. But this holds for each $\lambda \in \ell^\infty(G)$, and so $\Phi \,\Box\, \Psi = 0$. $\square$

We note that the proof of the above theorem would be significantly easier if we knew that Ω 0-clustered strongly on $S\times T$, for the conclusion would follow easily from equation (8.4) without any need for Theorem 3.3. Indeed, $\Phi \,\Box\, \Psi = 0$ $(\Phi \in B_\omega(S)\cap E_\omega^\circ, \Psi \in B_\omega(T)\cap E_\omega^\circ)$ whenever

$$\operatorname*{Lim}_{s\to\infty} \operatorname*{Lim\,sup}_{t\to\infty} \{\Omega(s,t) : s \in S,\, t\in T\} = 0\,.$$

However, we shall prove in Example 9.14 that there are weight functions ω on $\mathbb{Z}$ such that Ω 0-clusters on $\mathbb{N}\times\mathbb{N}$, but such that Ω does not 0-cluster strongly on $\mathbb{N}\times\mathbb{N}$.

COROLLARY 8.9. *Let ω be a weight on a group G, and let S be an infinite subset of G. Suppose that Ω 0-clusters on $S\times G$ and on $G\times S$. Then:*

(i) $\Phi\,\Box\,\Psi = 0$ *whenever* $\Phi\in B_\omega(S)\cap E_\omega^\circ$ *and* $\Psi\in E_\omega^\circ$*;*

(ii) $\Phi \diamond \Psi = 0$ *whenever* $\Phi\in E_\omega^\circ$ *and* $\Psi\in B_\omega(S)\cap E_\omega^\circ$*;*

(iii) $B_\omega(S)\subset \mathfrak{Z}_t^{(1)}(B_\omega)$, *and A_ω is not left strongly Arens irregular;*

(iv) $B_\omega(S)\cap E_\omega^\circ \subset R_\omega^\Box$.

PROOF. (i) Let $\Phi\in B_\omega(S)\cap E_\omega^\circ$ and $\Psi\in E_\omega^\circ$. By the theorem (with $T=G$), $\Phi\,\Box\,\Psi=0$.

(ii) This is similar.

(iii) This follows immediately from (i) and (ii).

(iv) This follows from Proposition 2.1 (with $I=E_\omega^\circ$) and the earlier results. □

COROLLARY 8.10. *Let ω be a weight on an abelian group G, and let S be an infinite subset of G. Suppose that Ω 0-clusters on $S\times G$. Then A_ω is not strongly Arens regular, and $R_\omega \neq \{0\}$.* □

THEOREM 8.11. *Let ω be a weight on a group G. Then the following conditions on ω are equivalent:*

(a) *the algebra $A_\omega = \ell^1(G,\omega)$ is Arens regular;*

(b) *the function Ω 0-clusters on $G\times G$;*

(c) $\Phi\,\Box\,\Psi = \Phi\diamond\Psi = 0$ *for each Φ and Ψ in E_ω°.*

PROOF. (c) $\Rightarrow$ (a) This is Proposition 2.16(i).

(a) $\Rightarrow$ (b) Assume towards a contradiction that (b) fails. Then there exist sequences (s_m) and (t_n) in G, each consisting of distinct points, such that $\lim_m \lim_n \Omega(s_m,t_n) = 2\delta$ for some $\delta>0$, say. By passing to a subsequence of (s_m), we may suppose that $\lim_n \Omega(s_m,t_n)>\delta$ for all $m\in\mathbb{N}$. As in [CrY, Theorem 1], we choose subsequences of (s_m)

and (t_n) inductively. Indeed, set $u_1 = s_1$, and let v_1 be the first element t_n with $\Omega(u_1, t_n) > \delta$. Having chosen $u_1, \dots, u_n$ and $v_1, \dots, v_n$, choose u_{n+1} to be the first element in the sequence (s_m) not in the set $\{u_i v_j v_k^{-1} : i, j, k \in \mathbb{N}_k\}$, and then choose v_{n+1} to be the first element in the sequence (t_n) which is not in the set $\{u_i^{-1} u_j v_k : i, j \in \mathbb{N}_{k+1},\ k \in \mathbb{N}\}$ and is such that $\Omega(u_i, v_{n+1}) > \delta$ $(i \in \mathbb{N}_{k+1})$. The sequences (u_m) and (v_n) are such that the elements $u_m v_n$ are all distinct for $m, n \in \mathbb{N}$ and $\Omega(u_m, v_n) > \delta$ whenever $m \le n$. Let f_m and g_n be the normalized point masses at u_m and v_n, respectively, and let λ be the characteristic function of the set $\{u_m v_n : m \le n\}$. By again passing to subsequences, if necessary, we may suppose that both repeated limits of $(\langle f_m \star g_n,\ \lambda\omega\rangle)$ exist; one is at least δ and the other is 0. Thus $\lambda\omega \circ m_A$ does not cluster on $(A_\omega)_{[1]} \times (A_\omega)_{[1]}$, and so A_ω is not Arens regular by Theorem 3.14, a contradiction of (a).

(b) $\Rightarrow$ (c) This follows from Corollary 8.9. □

Thus the algebra $\ell^1(G,\omega)$ is not Arens regular if and only if there exist sequences (s_m) and (t_n), each consisting of distinct elements of G, such that at least one of the two repeated limits

$$\lim_m \lim_n \Omega(s_m, t_n) \quad \text{and} \quad \lim_n \lim_m \Omega(s_m, t_n) \tag{8.5}$$

exists and is non-zero.

The equivalence of (a) and (b) in the above result is explicitly stated in [CrY, Theorem 1] and in [BaR, Corollary 3.8(i)]; in this latter paper, more general results, which apply when the group G is replaced by a semigroup, are given. However, the important equivalence with (c) seems to be new.

Note that the equivalence of (a) and (c) is a specific result applicable to Beurling algebras. For example, let A be a von Neumann algebra with predual E. Then A is Arens regular and E° is a closed $*$-ideal in the von Neumann algebra A''. For each Φ in $E^\circ \setminus \{0\}$, the element $\Phi^* \in E^\circ$, and $\Phi^* \,\Box\, \Phi \neq 0$.

The two repeated limits in (8.5) are zero if and only if the two repeated limits of the double sequence

$$((\delta^1\eta)(s_m, t_n) : m, n \in \mathbb{N}) \tag{8.6}$$

are equal to $+\infty$, where $\delta^1\eta$ was defined in (7.7).

Let ω be a weight on a countable group G such that ω is almost invariant and $1/\omega \in c_0$. Then clearly Ω 0-clusters on G. For example, we have the following specific result on $\mathbb{Z}$.

COROLLARY 8.12. *Let η be a subadddditive function on $\mathbb{Z}$ such that*

$$\lim_{j\to\infty} \eta(j) = \infty \quad \text{and} \quad \lim_{j\to\infty} s(j) = 0\,,$$

where s is the slope of η. Then $\ell^1(\exp \eta)$ is Arens regular.

PROOF. Since $\lim_{j\to\infty} s(j) = 0$, the corresponding function Ω 0-clusters on $\mathbb{Z} \times \mathbb{Z}$, and so this follows from Theorem 8.11. □

We also state the following corollary, specifically given as Corollary 1 in [CrY].

COROLLARY 8.13. (i) *Let G be a countable group. Then there is a weight ω on G such that $\ell^1(G,\omega)$ is Arens regular.*

(ii) *Let G be an uncountable group. Then there is no weight ω on G such that $\ell^1(G,\omega)$ is Arens regular.* □

At some points, we shall need an extension of the Theorem 8.11 to functions of more than two variables. We sketch the argument without giving full details.

Let ω be a weight on a group G, and let $k \in \mathbb{N}$ with $k \geq 2$. We set:

$$\Omega_k(s_1,\dots,s_k) = \frac{\omega(s_1\cdots s_k)}{\omega(s_1)\cdots\omega(s_k)} \quad (s_1,\dots,s_k \in G)\,. \tag{8.7}$$

In the first part of the next result, we restrict ourselves to abelian groups so that we can avoid some complicated expressions. In this case, the function Ω_k 0-clusters on $G \times \cdots \times G$ if and only if one fixed repeated limit is 0 whenever it exists.

THEOREM 8.14. *Let ω be a weight on a group G, and let $k \in \mathbb{N}$ with $k \geq 2$.*

(i) *Suppose that G is abelian. Then the following are equivalent:*

(a) *the function Ω_k 0-clusters on $G \times \cdots \times G$;*

(b) $\Phi_1 \,\square\, \cdots \,\square\, \Phi_k = 0$ *whenever* $\Phi_1,\dots,\Phi_k \in E_\omega^\circ$.

(ii) *Suppose that the function Ω_k 0-clusters strongly on $G \times \cdots \times G$. Then $\Phi_1 \,\square\, \cdots \,\square\, \Phi_k = \Phi_1 \diamond \cdots \diamond \Phi_k = 0$ whenever $\Phi_1,\dots,\Phi_k \in E_\omega^\circ$.*

PROOF. (i) (a) $\Rightarrow$ (b) By Proposition 3.6, the function Ω_k has an extension to a separately continuous function $\Omega_k : \beta G \times \cdots \times \beta G \to \mathbb{C}$, and further $\Omega_k(s_1^0,\dots,s_k^0) = 0$ whenever $s_1^0,\dots,s_k^0 \in \beta G \setminus G$. Take $\Phi_1,\dots,\Phi_k \in E_\omega^\circ$ and $\lambda \in \ell^\infty(G)$, and consider the expression

$$\begin{cases} \lim_{m_1} \lim_{m_2} \cdots \lim_{m_k} \\ \qquad \sum\{f_{m_1}(s_1) f_{m_2}(s_2)\cdots f_{m_k}(s_k)\lambda(s_1\cdots s_k)\Omega_k(s_1,\dots,s_k)\}\,, \end{cases} \tag{8.8}$$

the sum being taken over all elements $s_1, \ldots, s_k \in G$; here the functions $f_{m_1}, \ldots, f_{m_k} \in \ell^1(G)_{[1]}$. This repeated limit is equal to

$$\langle \Phi_1 \,\Box\, \cdots \,\Box\, \Phi_k,\ \lambda\omega \rangle\,.$$

The expression is analogous to that appearing in (8.4).

Fix $\varepsilon > 0$. Then there exist $(m_1(n), \ldots, m_k(n))$ in $\mathbb{N}^k$ for each $n \in \mathbb{N}$ such that each sequence $(m_j(n) : n \in \mathbb{N})$ consists of distinct points, and the appropriate sums in (8.8) are bounded by $\varepsilon \left\|\lambda\right\|_\infty$ for relevant combinations of the indices $m_1(n), \ldots, m_k(n)$. In this way, we see that the repeated limit in (8.8) is bounded by $\varepsilon \left\|\lambda\right\|_\infty$. This implies that (b) holds.

(b) $\Rightarrow$ (a) This is immediate.

(ii) This follows easily from equation (8.8). □

We now state a theorem that gives a condition for A_ω to be strongly Arens irregular. We do not prove the theorem at this stage because it is a special case of a more general theorem which will be proved in Theorem 11.9; see Corollary 11.10.

THEOREM 8.15. *Let G be a group, and let ω be a weight on G such that ω is diagonally bounded on S for some subset S of G with $|S| = |G|$. Then $\ell^1(G, \omega)$ is strongly Arens irregular.* □

COROLLARY 8.16. *Let ω be a weight on $\mathbb{Z}$. Suppose that there is strictly increasing sequence (n_k) in $\mathbb{N}$ such that*

$$\sup\,\{\omega(n_k)\omega(-n_k) : k \in \mathbb{N}\} < \infty\,.$$

Then $\ell^1(\mathbb{Z}, \omega)$ is strongly Arens irregular. □

The condition on ω in the above theorem is certainly satisfied if the weight ω is bounded on G. Example 9.3, below, will exhibit an easy weight ω on $\mathbb{Z}^2$ that is bounded on the subset $\{0\} \times \mathbb{Z}$, and so $\ell^1(\mathbb{Z}^2, \omega)$ is strongly Arens irregular, but such that ω is not diagonally bounded on the subset $\mathbb{Z} \times \{0\}$. Example 9.17 will exhibit a symmetric weight ω on $\mathbb{Z}$ that is diagonally bounded on an infinite subset S of $\mathbb{N}$, but which is unbounded. Further, Example 10.1 will show that, in the case where G is equal to $\mathbb{F}_2$, the free group on 2 generators, there is an unbounded weight ω which is diagonally bounded on the whole of G. These examples show that the above theorem is more general than those previously known.

Example 9.8 will show that the condition on ω in Theorem 8.15 cannot be replaced, in the case where $G = \mathbb{Z}$, by the weaker condition that $\{\omega(n) : n \in S\}$ be bounded for some infinite subset S of $\mathbb{Z}$, and

Example 9.16 will show that there is a weight ω on $\mathbb{Z}$, increasing on $\mathbb{Z}^+$, such that $\omega(-n) = 1$ $(n \in \mathbb{N})$, but nevertheless A_ω is not strongly Arens irregular.

A version of following ideas could be given for some more general groups G, but we restrict ourselves to the case where $G = \mathbb{Z}$.

DEFINITION 8.17. *Let ω be a weight on $\mathbb{Z}$. Then*

$$E_\omega^{\circ +} = E_\omega^\circ \cap B_\omega(\mathbb{Z}^+), \quad E_\omega^{\circ -} = E_\omega^\circ \cap B_\omega(\mathbb{Z}^-)\,.$$

Thus $E_\omega^{\circ -}$ and $E_\omega^{\circ -}$ are the 'parts of E_ω° at $+\infty$ and $-\infty$, respectively'. Clearly $E_\omega^{\circ -}$ and $E_\omega^{\circ -}$ are weak-$*$ closed subalgebras of B_ω, and

$$B_\omega = A_\omega \oplus E_\omega^{\circ -} \oplus E_\omega^{\circ -}\,. \tag{8.9}$$

THEOREM 8.18. *Let ω be a weight on $\mathbb{Z}$ such that*

$$\lim_{m\to\infty}\lim_{n\to\infty}\frac{\omega(m-n)}{\omega(m)\omega(-n)} = \lim_{m\to\infty}\lim_{n\to\infty}\frac{\omega(n-m)}{\omega(-m)\omega(n)} = 0\,. \tag{8.10}$$

(i) *Suppose that $\Phi \in E_\omega^{\circ -}$ and $\Psi \in E_\omega^{\circ +}$. Then $\Phi \,\Box\, \Psi = \Psi \,\Box\, \Phi = 0$.*

(ii) *The spaces $E_\omega^{\circ -}$ and $E_\omega^{\circ +}$ are both closed ideals in $(B_\omega, \,\Box\,)$.*

(iii) *Suppose that, for each $\Phi \in E_\omega^{\circ -} \setminus \{0\}$ (respectively, $E_\omega^{\circ +} \setminus \{0\}$), there exists $\Psi \in E_\omega^{\circ -}$ (respectively, $E_\omega^{\circ +}$) such that $\Phi \,\Box\, \Psi \neq \Psi \,\Box\, \Phi$. Then A_ω is strongly Arens irregular.*

(iv) *Suppose that there exist weights ω_- and ω_+ on $\mathbb{Z}$ such that*

$$\omega_- \mid \mathbb{Z}^- = \omega \mid \mathbb{Z}^- \quad \text{and} \quad \omega_+ \mid \mathbb{Z}^+ = \omega \mid \mathbb{Z}^+$$

and such that both ω_- and ω_+ are diagonally bounded on an infinite subset of $\mathbb{Z}$. Then A_ω is strongly Arens irregular.

PROOF. (i) It follows from condition (8.10) that Ω 0-clusters on $\mathbb{Z}^+ \times \mathbb{Z}^+$, and so this clause follows from Theorem 8.8.

(ii) Let $\Phi \in E_\omega^{\circ -}$. Clearly $f \,\Box\, \Phi, \Phi \,\Box\, f \in E_\omega^{\circ -}$ for each $f \in A_\omega$. We have remarked that $E_\omega^{\circ -}$ is a closed subalgebra of B_ω. Now let $\Psi \in E_\omega^{\circ +}$. Then $\Phi \,\Box\, \Psi = \Psi \,\Box\, \Phi = 0$ by (i). This shows that $E_\omega^{\circ -}$ is a closed ideal in B_ω. Similarly, $E_\omega^{\circ +}$ is a closed ideal in B_ω.

(iii) Take $\Phi \in B_\omega \setminus A_\omega$. Then Φ can be uniquely expressed in the form $\Phi = \Phi_- + \Phi_+$, where $\Phi_- \in E_\omega^{\circ -}$ and $\Phi_+ \in E_\omega^{\circ +}$; further, at least one of Φ_- and Φ_+ is not equal to 0, say $\Phi_+ \neq 0$. By hypothesis, there exists $\Psi \in E_\omega^{\circ +}$ with $\Phi_+ \,\Box\, \Psi \neq \Psi \,\Box\, \Phi_+$. By (i), $\Phi_- \,\Box\, \Psi = \Psi \,\Box\, \Phi_- = 0$, and so $\Phi \,\Box\, \Psi \neq \Psi \,\Box\, \Phi$. Hence $\Phi \notin \mathfrak{Z}(B_\omega)$. This shows that we have $\mathfrak{Z}(B_\omega) = A_\omega$, and so A_ω is strongly Arens irregular.

(iv) Take $\Phi \in E_\omega^{\circ +} \setminus \{0\}$. Then $\Phi \in E_{\omega_+}^{\circ +} \setminus \{0\}$. By Theorem 8.15, the algebra $\ell^1(\mathbb{Z}, \omega_+)$ is strongly Arens irregular, and so there exists $\Psi \in E_{\omega_+}^\circ$ with

$$\Phi \,\Box\, \Psi \neq \Psi \,\Box\, \Phi .$$

By (i), $\Psi \in E_{\omega_+}^{\circ +}$, and so $\Psi \in E_\omega^{\circ +} \subset B_\omega$. Thus $\Phi \notin \mathfrak{Z}(B_\omega)$. Similarily, $\Phi \notin \mathfrak{Z}(B_\omega)$ for each $\Phi \in E_\omega^{\circ -} \setminus \{0\}$. This again shows that $\mathfrak{Z}(B_\omega) = A_\omega$, and so A_ω is strongly Arens irregular. $\Box$

Let ω be a weight on $\mathbb{Z}$ such that ω is symmetric and $\omega \mid \mathbb{Z}^+$ is increasing and unbounded. Then ω satisfies condition (8.10).

Let ω be a weight on a group G. We have given necessary conditions and sufficient conditions for the strong Arens irregularity of the algebra A_ω. (The sufficient conditions were also obtained by Neufang [N5] using a different approach.) However we do not have the exact condition on ω equivalent to the strong Arens irregularity of A_ω, even in the special case where $G = \mathbb{Z}$. There are specific points that we cannot resolve. For example, let ω be a weight on $\mathbb{Z}$ such that

$$\liminf_{n\to\infty} \omega(n) < \infty \quad \text{and} \quad \liminf_{n\to\infty} \omega(-n) < \infty .$$

Does it follow that $\ell^1(\omega)$ is strongly Arens irregular?

We now seek information on the radicals $R_\omega^\Box$ and $R_\omega^\Diamond$ of B_ω. Our first result is an immediate corollary of Theorem 8.11.

THEOREM 8.19. *Let G be a group, and let ω be a weight on G such that A_ω is semisimple and Ω 0-clusters on $G \times G$. Then*

$$R_\omega^\Box = R_\omega^\Diamond = E_\omega^\circ \quad \text{and} \quad R_\omega^{\Box 2} = R_\omega^{\Diamond\, 2} = \{0\} .$$

$\Box$

The results that we have so far established suggest the following question.

Question Is it always true that $\mathfrak{Z}_t^{(1)}(B_\omega) \cap E_\omega^\circ \subset R_\omega^\Box$?

We now give a condition to ensure that $R_\omega^\Box \neq E_\omega^\circ$.

THEOREM 8.20. *Let G be a group, and let ω be a weight on G such that A_ω is semisimple. Suppose that there is a subsemigroup S of G and a sequence (t_n) consisting of distinct points of S such that*

$$\lim_{n\to\infty} \Omega(s, t_n) = 1 \quad (s \in S) . \tag{8.11}$$

Then $R_\omega^\Box \subsetneq E_\omega^\circ$.

PROOF. Since A_ω is semisimple, $R_\omega^\square \subset E_\omega^\circ$.

Consider the sequence $(\widetilde{\delta}_{t_n})$ in A_ω. We may suppose that there is a subnet (τ_β) of (t_n) such that $\lim_\beta \widetilde{\delta}_{\tau_\beta} = \Phi \in B_\omega$. Necessarily $\Phi \in E_\omega^\circ$ and $\|\Phi\| = 1$. Set $\lambda_0 = \chi_S\omega$, so that $\lambda_0 \in A'_\omega$ with $\|\lambda_0\| = 1$. Then

$$\langle \Phi, \lambda_0 \rangle = \lim_\beta \langle \widetilde{\delta}_{t_\beta}, \lambda_0 \rangle = 1\,.$$

We note that $\Phi \cdot \lambda_0 \in A'_\omega$ and that, for each $s \in S$, we have

$$\begin{aligned} \langle \delta_s, \Phi \cdot \lambda_0 \rangle &= \langle \Phi, \lambda_0 \cdot \delta_s \rangle = \lim_\beta \langle \widetilde{\delta}_{\tau_\beta}, \lambda_0 \cdot \delta_s \rangle = \lim_\beta \langle \delta_s \star \widetilde{\delta}_{\tau_\beta}, \lambda_0 \rangle \\ &= \lim_\beta \omega(s\tau_\beta)/\omega(\tau_\beta) = \lim_n \omega(st_n)/\omega(t_n) \\ &= \omega(s) = \lambda_0(s) \end{aligned}$$

by (8.11). It follows that $(\Phi \cdot \lambda_0) \mid S = \lambda_0 \mid S$.

We now *claim* that

$$\langle \Phi^{\square n}, \lambda_0 \rangle = \langle \Phi, \lambda_0 \rangle \quad (n \in \mathbb{N})\,. \tag{8.12}$$

To see this, first note that, for each $k \in \mathbb{N}$, we have $\langle \Phi^{\square k}, \mu \rangle = 0$ whenever $\mu \in A'_\omega$ and $\mu \mid S = 0$ because $t_{n_1} \cdots t_{n_k} \in S$ for each $n_1, \dots, n_k \in \mathbb{N}$. Equation (8.12) is trivial for $n = 1$. Now assume that (8.12) holds for $n = k$. Then

$$\begin{aligned} \langle \Phi^{\square (k+1)}, \lambda_0 \rangle &= \langle \Phi^{\square k}, \Phi \cdot \lambda_0 \rangle = \langle \Phi^{\square k}, (\Phi \cdot \lambda_0) \mid S \rangle \\ &= \langle \Phi^{\square k}, \lambda_0 \mid S \rangle = \langle \Phi^{\square k}, \lambda_0 \rangle = \langle \Phi, \lambda_0 \rangle\,, \end{aligned}$$

and so (8.12) holds for $n = k$. Thus the claim holds by induction.

It follows that

$$\|\Phi^{\square n}\| \geq |\langle \Phi, \lambda_0 \rangle| = 1 \quad (n \in \mathbb{N})\,,$$

and so the spectral radius of Φ in $(B_\omega, \square)$ is 1. Thus Φ is not quasi-nilpotent, and so, in particular, $\Phi \notin R_\omega^\square$. $\square$

Most of our earlier examples satisfy condition (8.11), above. For example, suppose that $G = \mathbb{Z}$, and let ω be any weight with

$$\omega(n) = \mathrm{e}^{an} \quad (n \in \mathbb{N})$$

for $a > 0$. Then ω satisfies (8.11) (with $S = \mathbb{N}$ and $t_n = n$ $(n \in \mathbb{N})$).

The following result was proved by Civin and Yood [CiY, Theorem 3.5] in the case where $\alpha = 0$, and a proof of the general case can be obtained by a modification of the earlier method. For details, see [La, pp. 41–47]. Let ω be a weight on $\mathbb{Z}$, and take $m \in \mathbb{N}$. An element $\Phi \in \mathcal{B}_\omega$ is *m-invariant* if

$$\langle \Phi, \delta_m \cdot \lambda \rangle = \langle \Phi, \lambda \rangle \quad (\lambda \in A'_\omega)\,.$$

Let $m \in \mathbb{N}$. The element $\lambda_m^r \in A_\omega$ is defined for $r \in \mathbb{Z}_{m-1}^+$ to have the value 1 at $n \in \mathbb{N}$ when $n \equiv r \bmod m$ and to be 0 otherwise. We set

$$I_m = \left\{\Phi \in B_\omega : \Phi \quad \text{is } m\text{−invariant}, \ \langle \Phi,\, \lambda_m^r \rangle = 0 \ (r \in \mathbb{Z}_{m-1}^+)\right\}.$$

THEOREM 8.21. *Let $\alpha \in \mathbb{R}^+$, and set $B_{\omega_\alpha} = (\ell^1(\mathbb{Z},\omega_\alpha)'',\,\Box\,)$. Then:*

(i) *for each $m \in \mathbb{N}$, there is an element of B_{ω_α} which is 2^{m+1}-invariant, but which is not 2^m-invariant;*

(ii) *for each $m \in \mathbb{N}$, I_m is a closed ideal in B_{ω_α} with $I_m \subset R_{\omega_\alpha}^\Box$;*

(iii) *the family $\{I_{2^m} : m \in \mathbb{N}\}$ is an ascending chain of distinct closed ideals in B_{ω_α}.*
In particular, $R_{\omega_\alpha}^\Box$ is infinite-dimensional.

Let G be a locally compact group. We have been seeking, inter alia, to show that the radicals $\mathcal{R}_\omega^\Box$ and $\mathcal{R}_\omega^\Diamond$ of the algebras $(\mathcal{B}_\omega,\,\Box\,)$ and $(\mathcal{B}_\omega,\,\Diamond\,)$ are non-zero. So far, we have three results in this direction.

First, suppose that $\mathcal{X}_\omega \neq \mathcal{A}'_\omega$. Then it follows from Theorem 5.4(iv) (with $X = \mathcal{X}_\omega = \mathcal{A}'_\omega \cdot \mathcal{A}_\omega$) that $\mathcal{X}_\omega^\circ$ is a non-zero, closed ideal contained in $\mathcal{R}_\omega^\Box$, and so the required result certainly holds in the case where G is not discrete. (For this remark in the case where $\omega = 1$, see [Gra3].) In fact, it is shown in [Gra3] that the radical of $(L^1(G)'',\,\Box\,)$ is not norm-separable for each non-discrete locally compact group G; the abelian case was proved earlier in [Gu].

Now suppose that the group G is infinite and discrete and that we are considering the radicals $R_\omega^\Box$ and $R_\omega^\Diamond$. Set $I = E_\omega^\circ$. In the case where Ω 0-clusters on $G\times G$, we see that $R_\omega^\Box = R_\omega^\Diamond = I$ by Theorem 8.19, and so $R_\omega^\Box$ and $R_\omega^\Diamond$ are non-zero. Again, suppose that A is commutative and that I is nilpotent of index $n \geq 2$ in $(A'',\,\Box\,)$. Then $I^{n-1} \subset R_\omega$, and so $R_\omega \neq \{0\}$. Finally, in the case where G is amenable and ω is an almost left-invariant weight on G, the closed ideal $R_\omega^\Box$ is non-zero by Theorem 7.40. Nevertheless, there are many natural weights on groups G that do not satisfy any of the above conditions; for example, this is the case for the weight ω on $\mathbb{Z}$ specified by $\omega(n) = \exp(|n|)$ $(n \in \mathbb{Z})$, as in (7.27). We now seek a 'hybrid' result that covers a wider class of weights, including the above one.

The ideas for finding elements in the radical of a second dual algebra go back at least to Civin and Yood [CiY], who in turn utilize formulae of Day from 1957 [Day].

DEFINITION 8.22. *Let ω be a weight on a group G, let S be a subset of G, and let $\lambda \in A_\omega(S)'$. Then*

$$I_\lambda(S) = \{\Lambda \in B_\omega(S) : s \cdot \Lambda = \lambda(s)\Lambda \ (s \in S),\ \langle \Lambda,\, \lambda\rangle = 0\}.$$

Clearly $I_\lambda(S)$ is always a weak-$*$ closed linear subspace of $B_\omega(S)$.

PROPOSITION 8.23. *Let ω be a weight on a group G, let S be a subsemigroup of G, and let $\lambda \in A_\omega(S)'$. Then $I_\lambda(S)$ is a closed left ideal in $(B_\omega(S), \Box)$, and $I_\lambda(S)$ is nilpotent of index 2.*

PROOF. Set $I = I_\lambda(S)$, and let $\Lambda \in I$ and $\Phi \in B_\omega(S)$. For each $s, t \in S$, we have

$$s \cdot (t \cdot \Lambda) = \lambda(t)\lambda(s)\Lambda = \lambda(s)t \cdot \Lambda .$$

It follows that $s \cdot (\Phi \Box \Lambda) = \lambda(s)\Phi \Box \Lambda$. Further, for each $f \in A_\omega(S)$, we have

$$f \cdot \Lambda = \left(\sum_{s\in S} f(s)\delta_s\right) \cdot \Lambda = \sum_{s\in S} f(s)\lambda(s)\Lambda = \langle f,\, \lambda\rangle\Lambda ,$$

and so $\Phi \Box \Lambda = \langle \Phi,\, \lambda\rangle\Lambda$. Hence

$$\langle \Phi \Box \Lambda,\, \lambda\rangle = \langle \Phi,\, \lambda\rangle\langle \Lambda,\, \lambda\rangle = 0 .$$

This shows that I is a left ideal in $B_\omega(S)$.

Clearly $\Lambda_1 \Box \Lambda_2 = \langle \Lambda_1,\, \lambda\rangle\Lambda_2 = 0 \;\; (\Lambda_1, \Lambda_2 \in I)$, and so I is nilpotent of index 2. $\Box$

We now seek a condition that ensures that $I_\lambda(S)$ is an ideal in B_ω, rather than just in $B_\omega(S)$.

DEFINITION 8.24. *Let S be a subsemigroup of a group G. Then S is* left thick *if, for each $t \in G$, there exists $s \in G$ such that $ts \in S$.*

Suppose that S is left thick and that $\{t_1, \dots, t_k\}$ is a finite subset of G. Successively choose $s_1, \dots, s_k \in S$ such that $t_i s_1 \cdots s_i \in S \;\; (i \in \mathbb{N}_k)$, and set $s = s_1 \cdots s_k \in S$. Then $t_i s \in G \;\; (i \in \mathbb{N}_k)$, and so S is left thick in the sense of [Pat, (1.20)]. Conversely, suppose that S is left thick in the latter sense, and take $t \in G$. Then there exists $s \in G$ with $\{ts, tts\} \subset S$, and then $ts \in S$ with $t(ts) \in S$, and so S is left thick in our sense. Thus our notion coincides with the classical one.

For example, for each $n \in \mathbb{N}$, $S = (\mathbb{Z}^+)^n$ is a left thick semigroup in $G = \mathbb{Z}^n$.

PROPOSITION 8.25. *Let ω be a weight on a group G, and let S be a subsemigroup of G. Suppose that S is left thick. Then, for each $\lambda \in A_\omega(S)'$, the set $I_\lambda(S)$ is a left ideal in $(B_\omega, \Box)$, and $I_\lambda(S) \subset R_\omega^\Box$.*

PROOF. Set $I = I_\lambda(S)$, and take $\Lambda \in I$. We know that

$$s \cdot \Lambda = \lambda(s)\Lambda \quad (s \in S).$$

Now take $t \in G$. Since S is left thick, there exists $u \in S$ such that $tu \in S$ and $t^{-1}u \in S$. For each $s \in S$, we have $stu \cdot \Lambda = \lambda(s)tu \cdot \Lambda$ because $tu \cdot \Lambda \in I$, and so

$$\lambda(u)st \cdot \Lambda = \lambda(u)\lambda(s)t \cdot \Lambda$$

because $u \cdot \Lambda = \lambda(u)\Lambda$. Thus $s \cdot (t \cdot \Lambda) = \lambda(s)t \cdot \Lambda$. Also

$$\lambda(t^{-1}u)\langle t \cdot \Lambda,\, \lambda\rangle = \langle u \cdot \Lambda,\, \lambda\rangle = \lambda(u)\langle \Lambda,\, \lambda\rangle = 0\,,$$

and so $\langle t \cdot \Lambda,\, \lambda\rangle = 0$. This shows that $t \cdot \Lambda \in I$.

It follows that $f \cdot \Lambda \in I$ for each $f \in A_\omega$. Since I is a weak-$\star$ closed, we have $\Phi \,\Box\, \Lambda \in I$ for each $\Phi \in B_\omega$, and so I is a left ideal in $(B_\omega, \,\Box\,)$.

We have noted in Proposition 8.23 that I is nilpotent, and so we have $I \subset R_\omega^\Box$. $\Box$

The next step is to exhibit cases where $I_\lambda(S) \neq 0$. Of course, this can only happen if the functional λ is *multiplicative* on S, in the sense that $\lambda(s_1 s_2) = \lambda(s_1)\lambda(s_2)$ $(s_1, s_2 \in S)$.

DEFINITION 8.26. *Let ω be a weight on a semigroup S. Then ω is* almost multiplicative on S *if*

$$\operatorname*{Lim}_{s\to\infty}\operatorname*{Lim}_{t\to\infty} \frac{\omega(st)}{\omega(s)\omega(t)} = 1\,,$$

where the limits are taken with $s, t \in S$.

THEOREM 8.27. *Let ω be a weight on an infinite, amenable group G, and let S be a subsemigroup of G. Suppose that S is left thick and that ω is almost multiplicative on S. Then*

$$\dim R_\omega^\Box \geq 2^{2^{\aleph_0}}\,.$$

PROOF. Set $A_\omega = \ell^1(G,\omega)$, and set $\mathfrak{m} = 2^{2^{\aleph_0}}$. For each $s \in S$, set

$$\lambda_0(s) = \operatorname*{Lim}_{t\to\infty} \frac{\omega(st)}{\omega(t)}\,,$$

where the limit is taken for $t \in S$, and the limit exists because ω is almost multiplicative on S. Then $\lambda_0 \in A_\omega(S)'$, λ_0 is mutiplicative on S, and

$$\operatorname*{Lim}_{s\to\infty} \lambda_0(s)/\omega(s) = 1;$$

we may suppose that $\lambda_0(s)/\omega(s) \geq 1/2$ $(s \in S)$.

Set $I = I_{\lambda_0}(S)$. By Proposition 8.25, $I \subset R_\omega^\Box$, and so it suffices to show that $\dim I \geq \mathfrak{m}$.

Let M be an element of $\ell^1(S)'' \subset \ell^1(G)''$ such that

$$s \cdot \mathrm{M} = \mathrm{M} \quad (s \in S) \quad \text{and} \quad \langle \mathrm{M}, \chi_S \rangle = 1\,.$$

Then $t \cdot \mathrm{M} = \mathrm{M}$ $(t \in G)$, and so $\mathrm{M} \mid c_0(S) = 0$.

Define $\mathrm{N} \in B_\omega$ by

$$\langle \mathrm{N}, \lambda \rangle = \langle \mathrm{M}, \lambda/\lambda_0 \rangle \quad (\lambda \in A_\omega')\,;$$

we note that in fact $\|\mathrm{N}\| = \|\mathrm{M}\|$. Also $\langle \mathrm{N}, \lambda_0 \rangle = 1$.

Take $\lambda \in A_\omega(S)'$ and $s \in S$, and set

$$\mu = \lambda_0(s)(\lambda/\lambda_0) \cdot \delta_s - (\lambda \cdot \delta_s)/\lambda_0\,.$$

For each $t \in S$, we have

$$|\mu(t)| = \left| \frac{\lambda_0(s)\lambda(st)}{\lambda_0(st)} - \frac{\lambda(st)}{\lambda_0(t)} \right| \leq 2\,\|\lambda\| \left| \lambda_0(s) - \frac{\lambda_0(st)}{\lambda_0(t)} \right|,$$

and so $\mu \mid S \in c_0(S)$. Thus $\langle \mathrm{M}, \mu \rangle = 0$. It follows that $s \cdot \mathrm{N} = \lambda_0(s)\mathrm{N}$.

By [Mi1, Theorem 9], the semigroup S is left amenable (see also [Pat, Proposition (1.21)]), and so, by [Pat, Theorem (7.8)], the set of elements M in $\ell^1(S)''$ such that $s \cdot \mathrm{M} = \mathrm{M}$ $(s \in S)$ and $\langle \mathrm{M}, \chi_S \rangle = 1$ has cardinality at least $\mathfrak{m}$. The corresponding set of elements N in $B_\omega(S)$ has the same cardinality, and the difference of any two distinct elements in this latter set belongs to I.

The result follows. $\Box$

We shall give in the next chapter examples of weights on $\mathbb{Z}$ that satisfy the conditions of the above theorem.

Unfortunately, several obvious questions about the radicals of the second duals of group and Beurling algebras remain open. Set

$$B_{\omega_\alpha} = (\ell^1(\omega_\alpha)'', \Box)\,,$$

and take $\{I_m : m \in \mathbb{N}\}$ to be the family of closed ideals of B_{ω_α} that was discussed in Theorem 8.21. Now set

$$I = \overline{\bigcup \{I_m : m \in \mathbb{N}\}}\,.$$

Then I is 'large' closed left ideal contained in $R_{\omega_\alpha}^\Box$, and $I^{\Box 2} = 0$. We do not know whether or not $I = R_{\omega_\alpha}^\Box$, and whether or not $R_{\omega_\alpha}^\Box$ is itself nilpotent of index 2. Let G be a group, and let $R^\Box$ and $R^\Diamond$ be the radicals of the two second duals of $\ell^1(G)$. Then we do not know whether or not $R^\Box$ and $R^\Diamond$ are nilpotent, and whether or not they are always equal; we guess that the latter is not the case.

CHAPTER 9

Algebras on Discrete, Abelian Groups

In the next two chapters, we shall present a variety of examples which show the limits of our earlier theorems. In the present chapter all our examples will be commutative, and almost all will be of the form $\ell^1(\omega) = \ell^1(\mathbb{Z},\omega)$ for a suitably chosen weight ω on $\mathbb{Z}$. Again set $A_\omega = \ell^1(\omega)$ and $B_\omega = A''_\omega$, etc. Thus, for all the examples in this chapter, there is just one topological centre, equal to the centre $\mathfrak{Z}(B_\omega)$ of $(B_\omega, \Box)$.

The first example was essentially already given by Craw and Young in [CrY]. In the example, $\|\cdot\|$ denotes the Euclidean norm on $\mathbb{Z}^k$.

EXAMPLE 9.1. *Fix $k \in \mathbb{N}$, and consider the group $(\mathbb{Z}^k, +)$. Let*

$$\omega_\alpha(n) = (1 + \|n\|)^\alpha \quad (n \in \mathbb{Z}^k),$$

where $\alpha > 0$. Then $\ell^1(\mathbb{Z}^k, \omega_\alpha)$ is Arens regular, and the radical R_{ω_α} of $(\ell^1(\mathbb{Z}^k, \omega_\alpha)'', \Box)$ is equal to $E^\circ_{\omega_\alpha}$.

PROOF. The corresponding function Ω_α 0-clusters on $\mathbb{Z}^k \times \mathbb{Z}^k$, and so this follows from Theorem 8.11 and Theorem 8.19. $\Box$

Take $\alpha > 0$. Since the Banach algebra $\ell^1(\mathbb{Z}, \omega_\alpha)$ is Arens regular, we have $WAP(\ell^1(\mathbb{Z}, \omega_\alpha)) = \ell^\infty(\mathbb{Z}, 1/\omega_\alpha)$. However, it is not the case that

$$WAP(\mathbb{Z}, 1/\omega_\alpha) = \ell^\infty(\mathbb{Z}, 1/\omega_\alpha).$$

Indeed, take $\lambda \in \ell^\infty(\mathbb{Z}) \setminus WAP(\mathbb{Z})$: we have $\lambda\omega_\alpha \in \ell^\infty(\mathbb{Z}, 1/\omega_\alpha)$, but $\lambda\omega_\alpha \notin WAP(\mathbb{Z}, 1/\omega_\alpha)$. Thus

$$WAP(\ell^1(\mathbb{Z}, \omega_\alpha)) \subsetneq WAP(\mathbb{Z}, 1/\omega_\alpha). \tag{9.1}$$

EXAMPLE 9.2. *There is a sequence (A_k) of Arens regular Banach algebras such that $c_0(A_k)$ is Arens regular, but $\ell^\infty(A_k)$ is not Arens regular.*

PROOF. For each $k \in \mathbb{N}$, set $A_k = \ell^1(\omega_{1/k})$, so that, by Example 9.1, A_k is an Arens regular Banach algebra.

Set

$$\mathfrak{A} = \ell^\infty(A_k) = \left\{a = (a_k) \in \prod A_k : \|a\| = \sup \|a_k\| < \infty\right\},$$

so that $\mathfrak{A}$ is a commutative Banach algebra.

For each $k \in \mathbb{N}$, we define $\lambda_k \in (A_k')_{[1]}$ to be the map

$$\lambda_k : \sum_{j=-\infty}^{\infty} \alpha_j \delta_j \mapsto \sum_{j=0}^{\infty} \alpha_j \omega_{1/k}(j)\,.$$

Let $\mathcal{U}$ be an ultrafilter on $\mathbb{N}$, and define λ on $\mathfrak{A}$ by

$$\langle a,\, \lambda \rangle = \lim_{\mathcal{U}} \langle a_k,\, \lambda_k \rangle\,;$$

the limit always exists, and $\lambda \in \mathfrak{A}'$ with $\|\lambda\| = 1$.

For each $m, n \in \mathbb{N}$, define $a_m = (a_{m,k})$ and $b_n = (b_{n,k})$ by setting

$$a_{m,k} = \delta_m / \omega_{1/k}(m), \quad b_{n,k} = \delta_{-n} / \omega_{1/k}(n), \quad (k \in \mathbb{N})\,.$$

For each $m, n, k \in \mathbb{N}$, we have $a_{m,k}, b_{n,k} \in A_k$ with $\|a_{m,k}\| = \|b_{n,k}\| = 1$, and so $a_m, b_n \in \mathfrak{A}$. For $n > m$, we have $\langle a_{m,k} \star b_{n,k},\, \lambda_k \rangle = 0$ because $m - n \notin \mathbb{Z}^+$, and so

$$\lim_m \lim_n \langle a_m b_n,\, \lambda \rangle = 0\,.$$

For $m > n$, we have

$$\langle a_{m,k} \star b_{n,k},\, \lambda_k \rangle = \frac{\omega_{1/k}(m-n)}{\omega_{1/k}(m)\omega_{1/k}(n)} = \left(\frac{1 + |m-n|}{(1+|m|)(1+|n|)} \right)^{1/k} \to 1$$

as $k \to \infty$, and so

$$\lim_n \lim_m \langle a_m b_n,\, \lambda \rangle = 1\,.$$

It follows from (3.2) that λ is not weakly almost periodic, and so $\mathfrak{A}$ is not Arens regular.

Set $B = c_0(A_k)$. Then B is a Banach algebra which is Arens regular by a remark in Chapter 2 on page 35. However $\ell^\infty(B)$ contains the algebra $\mathfrak{A} = \ell^\infty(A_k)$ as a closed subalgebra, and so $\ell^\infty(B)$ is not Arens regular. □

EXAMPLE 9.3. *There is a Beurling algebra which is strongly Arens irregular and which has as a closed subalgebra an infinite-dimensional Beurling algebra which is Arens regular.*

PROOF. Let

$$\omega(m,n) = (1 + |m|)^\alpha \quad (m, n \in \mathbb{Z})\,,$$

where $\alpha > 0$. Then ω is a weight on $(\mathbb{Z}^2, +)$. Set $A_\omega = \ell^1(\mathbb{Z}^2, \omega)$.

Clearly ω is bounded, and hence diagonally bounded, on the set $S = \{0\} \times \mathbb{Z}$, and so it follows from Theorem 8.15 that the algebra A_ω is strongly Arens irregular.

Let $B = \{f \in A_\omega : \operatorname{supp} f \subset \mathbb{Z} \times \{0\}\}$, so that B is a closed subalgebra of A_ω. Then B is identified with the Banach algebra $\ell^1(\mathbb{Z}, \omega_\alpha)$, which was specified in Example 9.1, and so B is Arens regular.

Thus $B'' \cap \mathfrak{Z}(A'') = B$, whereas $\mathfrak{Z}(B'') = B''$, and this shows that the inclusion in (2.23) can be strict. □

EXAMPLE 9.4. *Let $\omega = \omega_\alpha \otimes \omega_\beta$ on $\mathbb{Z}^2$, so that*

$$\omega(m, n) = (1 + |m|)^\alpha (1 + |n|)^\beta \quad (m, n \in \mathbb{Z}),$$

where $\alpha, \beta > 0$. Then ω is a weight on $(\mathbb{Z}^2, +)$. The algebra $\ell^1(\mathbb{Z}^2, \omega)$ is neither Arens regular nor strongly Arens irregular.

PROOF. Certainly ω is a weight on $\mathbb{Z}^2$. Set $A_\omega = \ell^1(\mathbb{Z}^2, \omega)$.

Let $s_m = (m, 0)$ $(m \in \mathbb{N})$ and $t_n = (0, n)$ $(n \in \mathbb{N})$. Then (s_m) and (t_n) are sequences in $\mathbb{Z}^2$, each consisting of distinct points. Clearly we have $\Omega(s_m, t_n) = 1$ $(m, n \in \mathbb{N})$, and so Ω does not 0-cluster on $\mathbb{Z}^2 \times \mathbb{Z}^2$. By Theorem 8.11, A_ω is not Arens regular.

Let $S = \{(m, m) : m \in \mathbb{N}\}$, so that S is an infinite subsemigroup of $\mathbb{Z}^2$.

Let $x_m = (m, m) \in S$, and let (y_n) be a sequence of distinct points in $\mathbb{Z}^2$ such that the two repeated limits of $(\Omega(x_m, y_n) : m, n \in \mathbb{Z})$ both exist. We have

$$\lim_m \Omega(x_m, (r, s)) = 1/(1 + |r|)^\alpha (1 + |s|)^\beta \quad (r, s \in \mathbb{Z}),$$

and so $\lim_n \lim_m \Omega(x_m, y_n) = 0$. Also

$$\lim_n \Omega(x_m, y_n) \leq 1/(1 + m)^{2\alpha} \quad (m \in \mathbb{N}),$$

and so $\lim_m \lim_n \Omega(x_m, y_n) = 0$. Thus Ω 0-clusters on $S \times \mathbb{Z}^2$. It follows from Corollary 8.9(iii) that $B_\omega(S) \subset \mathfrak{Z}(B_\omega)$, and so A_ω is not strongly Arens irregular. □

EXAMPLE 9.5. (i) *Let ω be a weight on $\mathbb{Q}$ such that ω is bounded on the set $\{r \in \mathbb{Q} : |r| \leq 1\}$. Then $\ell^1(\mathbb{Q}, \omega)$ is strongly Arens irregular.*

(ii) *There is a weight on ω on $\mathbb{Q}$ such that $\ell^1(\mathbb{Q}, \omega)$ is Arens regular.*

PROOF. (i) The weight ω is diagonally bounded on the set

$$\{r \in \mathbb{Q} : |r| \leq 1\},$$

and so the result follows from Theorem 8.15.

(ii) Such a weight ω is constructed in [CrY, Corollary 1]. □

EXAMPLE 9.6. *Take* ω *be a weight on* $\mathbb{Z}$ *such that*

$$\omega(n) = \mathrm{e}^{an}, \quad \omega(-n) = \mathrm{e}^{bn} \quad (n \text{ sufficiently large in } \mathbb{Z}^+),$$

where $a, b \geq 0$ *and* $\max\{a, b\} > 0$. *Then* ω *is unbounded, but* $\ell^1(\mathbb{Z}, \omega)$ *is strongly Arens irregular. Further,*

$$\{0\} \neq R_\omega \subsetneq E_\omega^\circ ,$$

and so $R_\omega \not\subset \mathfrak{Z}(B_\omega)$.

PROOF. The weight ω satisfies equation (8.10): for example,

$$\omega(m-n)/\omega(m)\omega(-n) = \exp(b(n-m) - am - bn) = \exp(-(a+b)m)$$

for $m, n \in \mathbb{Z}^+$ with $m < n$, and so

$$\lim_{m\to\infty} \lim_{n\to\infty} \frac{\omega(m-n)}{\omega(m)\omega(-n)} = 0 .$$

Also, there are obvious weights ω_- and ω_+ on $\mathbb{Z}$ with $\omega_- \mid \mathbb{Z}^- = \omega \mid \mathbb{Z}^-$ and $\omega_+ \mid \mathbb{Z}^+ = \omega \mid \mathbb{Z}^+$ and such that ω_- and ω_+ are diagonally bounded on $\mathbb{Z}$, and so $\ell^1(\mathbb{Z}, \omega)$ is strongly Arens irregular by Theorem 8.18(iv).

By Theorem 8.20 (with $S = \mathbb{Z}^+$ and $t_n = n$ for each $n \in \mathbb{N}$), we have $R_\omega \subsetneq E_\omega^\circ$. The weight ω is almost multiplicative on $\mathbb{Z}^+$, and so it follows from Theorem 8.27 that $R_\omega \neq \{0\}$. □

Examples of the above form show that, given any weight ω on $\mathbb{Z}$, there is a weight $\overline{\omega}$ on $\mathbb{Z}$ such that $\ell^1(\overline{\omega}) \subset \ell^1(\omega)$ and $\ell^1(\overline{\omega})$ is strongly Arens irregular.

EXAMPLE 9.7. *There is a weight* ω *on* $\mathbb{Z}$ *such that the Banach algebra* $\ell^1(\mathbb{Z}, \omega)$ *is neither Arens regular nor strongly Arens irregular. Further, we have*

$$\{0\} \neq R_\omega \subsetneq E_\omega^\circ .$$

PROOF. We define $\eta : \mathbb{Z} \to \mathbb{R}^+$ by setting

$$\eta(-j) = j, \quad \eta(j) = \log(1 + j) \quad (j \in \mathbb{Z}^+) .$$

It is easily checked that η is subadditive on $\mathbb{Z}$; also $\eta(0) = 0$. A graph of the function η, regarded as a function on $\mathbb{R}$, is indicated in Figure 1 on page 115.

Set $\omega = \exp \eta$ and $A_\omega = \ell^1(\mathbb{Z}, \omega)$.

We first *claim* that A_ω is not Arens regular. Indeed,

$$(\delta^1\eta)(-j, -k) = 0 \quad (j, k \in \mathbb{N}) ,$$

and so both repeated limits of $(\Omega(-j, -k) \, : \, j, k \in \mathbb{N})$ are 1. Thus Ω does not 0-cluster on $\mathbb{Z} \times \mathbb{Z}$, and so, by Theorem 8.11, A_ω is not Arens regular.

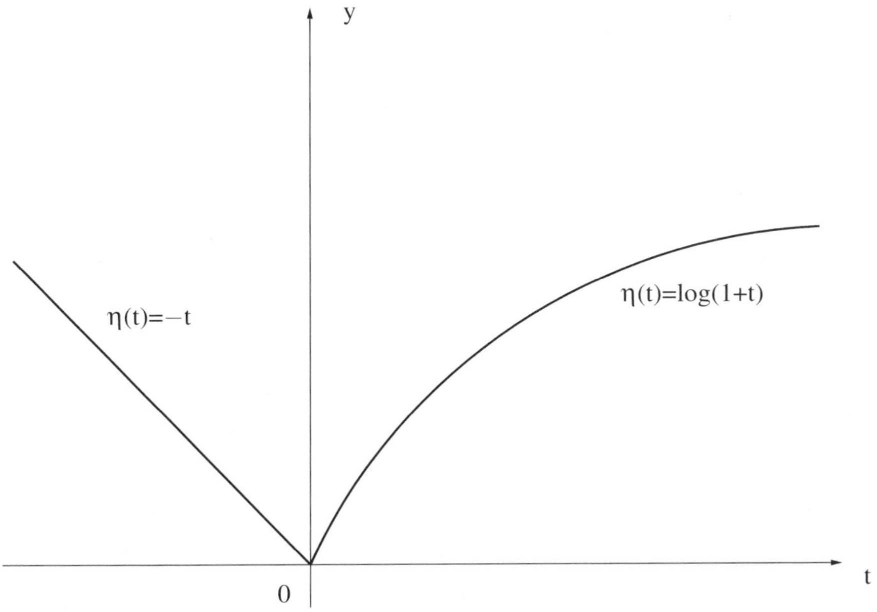

FIGURE 1

Set

$$\alpha_{j,k} = (\delta^1\eta)(j,k), \quad \beta_{j,k} = (\delta^1\eta)(j,-k) \quad (j,k \in \mathbb{N}),$$

so that $\beta_{j,k} = \alpha_{j,-k}$ $(j,k \in \mathbb{N})$. We shall show that

$$(9.2) \qquad \begin{cases} \lim_j \liminf_k \alpha_{j,k} = \infty\,, & \lim_k \liminf_j \alpha_{j,k} = \infty\,, \\ \lim_j \liminf_k \beta_{j,k} = \infty\,, & \lim_k \liminf_j \beta_{j,k} = \infty\,; \end{cases}$$

it will then follow that Ω 0-clusters on $\mathbb{N} \times \mathbb{Z}$. By Corollary 8.10, this is sufficient to imply that A_ω is not strongly Arens irregular and that $R_\omega \neq \{0\}$.

Throughout, $j,k \in \mathbb{N}$.

First, we have

$$\alpha_{j,k} = \log(1+j) + \log(1+k) - \log(1+j+k) \geq \log(1+j) - j/(1+k),$$

and so $\liminf_k \alpha_{j,k} \geq \log(1+j)$ for each j, whence

$$\lim_j \liminf_k \alpha_{j,k} = \infty\,.$$

By symmetry, $\lim_k \liminf_j \alpha_{j,k} = \infty$.

Second, for each j and each $k > j$, we have

$$\beta_{j,k} = \eta(j) + k - (k-j) \geq j\,,$$

and so $\liminf_k \beta_{j,k} \geq j$ for each j, whence

$$\lim_j \liminf_k \beta_{j,k} = \infty .$$

Also, for each k and each $j > k$, we have $\eta(j-k) \leq \eta(j)$, and so it follows that $\beta_{j,k} \geq \eta(-k) = k$. Thus $\liminf_j \beta_{j,k} \geq k$ for each k, whence

$$\lim_k \lim_j \beta_{j,k} = \infty .$$

Thus conditions (9.2) hold, as required.

By Theorem 8.20 (with $S = \mathbb{Z}^-$ and $t_n = -n$ for each $n \in \mathbb{N}$), we also have $R_\omega \subsetneq E_\omega^\circ$. □

The next example is similar, but a little more complicated. It shows that an obvious conjecture on the strong Arens irregularity of the algebras A_ω is false.

EXAMPLE 9.8. *There is a weight ω on $\mathbb{Z}$ such that Ω does not cluster on $\mathbb{Z} \times \mathbb{Z}$, such that*

$$\limsup_{n\to\infty} \omega(n) = \infty \quad \text{and} \quad \liminf_{n\to\infty} \omega(n) < \infty ,$$

and such that the Banach algebra $\ell^1(\mathbb{Z}, \omega)$ is neither Arens regular nor strongly Arens irregular. Further, $\{0\} \neq R_\omega \subsetneq E_\omega^\circ$.

PROOF. We again define a function η on $\mathbb{Z}$ (and then set $\omega = \exp \eta$ and $A_\omega = \ell^1(\mathbb{Z}, \omega)$).

As before, we set

$$\eta(-j) = j \quad (j \in \mathbb{Z}^+) ,$$

and this is sufficient to show that A_ω is not Arens regular and that $R_\omega \neq \{0\}$.

It remains to define η on $\mathbb{Z}^+$. We shall specify two strictly increasing sequences (m_k) and (n_k) in $\mathbb{N}$ such that

$$2 = m_1 < n_1 < m_2 < n_2 < \cdots < m_k < n_k < \cdots .$$

For convenience, we shall then set $\gamma_k = 1/n_k$ $(k \in \mathbb{N})$, so that (γ_k) is a strictly decreasing sequence in the interval $(0,1) \cap \mathbb{Q}$ such that $\gamma_k \to 0$ as $k \to \infty$.

First, $\eta(j) = j$ $(j = 0, 1, 2)$. Now suppose that $\eta(j)$ has been defined on $\mathbb{N}_{m_k}$ such that $\eta(m_k) \in \mathbb{N}$, $\eta(m_k) \geq k$, and

$$\eta(j) \leq \eta(m_k) \quad (j \in \mathbb{N}_{m_k}) .$$

We choose

$$n_k = m_k + \eta(m_k) - 1 ,$$

and then define

$$\eta(m_k + j) = \eta(m_k) - j \quad (j \in \mathbb{N}_{n_k - m_k}).$$

Note that $\eta(n_k) = 1$ and that $n_k - m_k \to \infty$ as $k \to \infty$. Once we have fixed m_k and n_k, we choose $m_{k+1} \in \mathbb{N}$ so that $m_{k+1} > n_k$, so that $\gamma_k m_{k+1} \in \mathbb{N}$, and so that $\gamma_k m_{k+1} > \max\{k + 1, \eta(m_k)\}$; this choice is clearly possible. We now define

$$\eta(j) = \gamma_k j \quad (j = n_k, \ldots, m_{k+1}).$$

This provides an inductive construction of the sequences (m_k) and (n_k) and of η on $\mathbb{N}$.

A graph of the function η, regarded as a function on $\mathbb{R}^+$, is indicated in Figure 2, below.

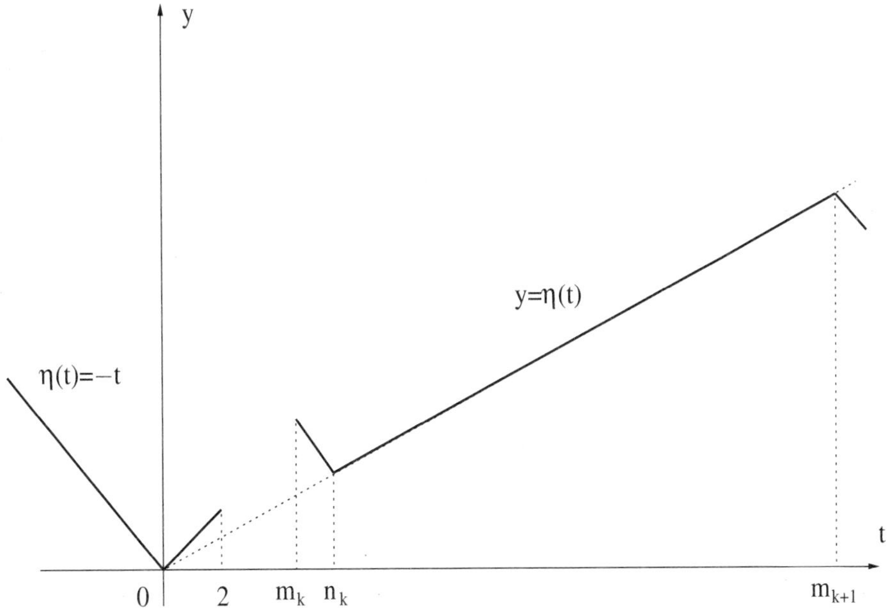

FIGURE 2

We notice that the function η on $\mathbb{N}$ has the following properties. First, for each $k \in \mathbb{N}$, we have

$$\eta(j) \geq \gamma_k j \quad (j \in \mathbb{N}_{m_{k+1}}) \quad \text{and} \quad \eta(j) \leq \gamma_k j \quad (j \geq n_k).$$

Second, $\eta(m_k) > k$ $(k \in \mathbb{N})$ and $(\eta(m_k))$ is a strictly increasing sequence with $\lim_k \eta(m_k) = \infty$, so that $\limsup_n \omega(n) = \infty$. Third, we have arranged that $\eta(n_k) = 1$ $(k \in \mathbb{N})$, and so $\liminf_n \omega(n) = \mathrm{e} < \infty$, one of the requirements of the example. Finally, for each $k \in \mathbb{N}$, we have

$$\eta(m_k) = \max\{\eta(j) : j \in \mathbb{N}_{m_k}\}.$$

As before, the slope of the function η is defined to be s, where

$$s(j) = \eta(j+1) - \eta(j) \quad (j \in \mathbb{N}).$$

Clearly $s(j) \geq -1$ $(j \in \mathbb{N})$ and $s(j) \leq \gamma_k$ for $j \geq n_k$, and so we have $\limsup_j s(j) = 0$.

We *claim* that the function η is subadditive on $\mathbb{Z}$: we require that

$$\eta(m+n) \leq \eta(m) + \eta(n) \quad (m, n \in \mathbb{Z}). \tag{9.3}$$

The proof is by induction. The result is trivially true when $m, n \in \mathbb{N}$ and $m + n \in \mathbb{Z}^+_{m_1}$. Now assume that (9.3) holds whenever $m, n \in \mathbb{N}$ and $m + n \in \mathbb{Z}^+_{m_k}$, where $k \in \mathbb{N}$. Take $m, n \in \mathbb{N}$ with $m + n \in \mathbb{Z}^+_{m_{k+1}}$. First, suppose that $m_k \leq m + n \leq n_k$. If $m, n \leq m_k$, then

$$\eta(m+n) \leq \eta(m_k) = \gamma_{k-1} m_k \leq \gamma_{k-1} m + \gamma_{k-1} n \leq \eta(m) + \eta(n).$$

If $m \geq m_k$, then

$$\eta(m+n) - \eta(m) = -n \leq \eta(n),$$

and similarly if $n \geq m_k$. Second, suppose that $n_k \leq m + n \leq m_{k+1}$. Then

$$\eta(m+n) = \gamma_k m + \gamma_k n \leq \eta(m) + \eta(n).$$

Thus (9.3) holds whenever $m, n \in \mathbb{N}$ and $m + n \in \mathbb{Z}^+_{m_{k+1}}$. It follows that inequality (9.3) holds whenever $m, n \in \mathbb{N}$.

Certainly, (9.3) holds whenever $-m, -n \in \mathbb{N}$, and so it remains to show that

$$\eta(m-n) \leq \eta(m) + \eta(-n) = \eta(m) - n \quad (m, n \in \mathbb{N}). \tag{9.4}$$

But now $\eta(m) \geq -n + \eta(m-n)$ for $m, n \in \mathbb{N}$ because the slope of η is bounded below by -1, and so (9.4) follows.

We have established the claim that η is subadditive.

We now show that Ω does not cluster on $\mathbb{Z} \times \mathbb{Z}$. Indeed, first fix $n \in \mathbb{N}$. For each sufficiently large k, we have $n_k - n > m_k$, and so

$$\eta(n_k - n) - \eta(n_k) - \eta(-n) = 0.$$

Thus $\lim_k \Omega(n_k, -n) = 1$, and so $\lim_n \lim_k \Omega(n_k, -n) = 1$. Second, fix $k \in \mathbb{N}$. For each sufficiently large n, we have $n_k - n < 0$, and so

$$\eta(n_k - n) - \eta(n_k) - \eta(-n) = n - n_k - 1 - n = -n_k - 1.$$

Hence $\lim_k \lim_n \Omega(n_k, -n) = 0$. Thus Ω does not cluster on $\mathbb{Z} \times \mathbb{Z}$.

We take $S = \{m_k : k \in \mathbb{N}\}$, an infinite subset of $\mathbb{Z}$. We now define

$$\alpha_{j,k} = (\delta^1\eta)(m_k, j), \quad \beta_{j,k} = (\delta^1\eta)(m_k, -j) \quad (j, k \in \mathbb{N});$$

we shall again show that

$$(9.5)\qquad \begin{cases} \lim_j \liminf_k \alpha_{j,k} = \infty\,, & \lim_k \liminf_j \alpha_{j,k} = \infty\,, \\ \lim_j \liminf_k \beta_{j,k} = \infty\,, & \lim_k \liminf_j \beta_{j,k} = \infty\,. \end{cases}$$

This implies that Ω 0-clusters on $S \times \mathbb{Z}$, and hence, by Corollary 8.10, A_ω is not strongly Arens irregular and $R_\omega \neq \{0\}$.

First, fix $j \in \mathbb{N}$. For k sufficiently large in $\mathbb{N}$, the interval $[m_k, n_k,]$ has length $n_k - m_k \geq j$, and so $\eta(m_k + j) = \eta(m_k) - j$. Thus

$$\alpha_{j,k} = \eta(j) + j\,,$$

and so $\liminf_k \alpha_{j,k} = \eta(j) + j \geq j$. Hence $\lim_j \lim_k \alpha_{j,k} = \infty$.

Second, fix $k \in \mathbb{N}$, and set $\varepsilon = \eta(m_k)/2m_k$, so that $\varepsilon > 0$. Since $\limsup_j s(j) = 0$, there exists $j_0 \in \mathbb{N}$ such that $s(j) < \varepsilon$ $(j \geq j_0)$, and this implies that

$$\eta(m_k + j) - \eta(j) \leq m_k s(j) < \varepsilon m_k \quad (j \geq j_0)\,.$$

Thus

$$\alpha_{j,k} \geq \eta(m_k) - \varepsilon m_k \geq \eta(m_k)/2 \quad (j \geq j_0)\,,$$

and so $\liminf_j \alpha_{j,k} \geq \eta(m_k)/2$. Since $\lim_k \eta(m_k) = \infty$, this shows that $\lim_k \liminf_j \alpha_{j,k} = \infty$.

Third, fix $j \in \mathbb{N}$. For k sufficiently large in $\mathbb{N}$, we have $m_k > j$, and so $0 \leq m_k - j \leq m_k$ and $\eta(m_k - j) \leq \eta(m_k)$. Thus $\beta_{j,k} \geq \eta(-j) = j$, and so $\liminf_k \beta_{j,k} \geq j$. Hence $\lim_j \liminf_k \beta_{j,k} = \infty$.

Finally, again fix $k \in \mathbb{N}$. For j sufficiently large in $\mathbb{N}$, we have $j > m_k$, and so $\eta(-j) - \eta(m_k - j) = m_k$. Thus $\beta_{j,k} \geq m_k$, and so $\liminf_j \beta_{j,k} \geq m_k$. Hence $\lim_k \liminf_j \beta_{j,k} = \infty$.

We have shown that conditions (9.5) hold, as required.

Again by Theorem 8.20 (with $S = \mathbb{Z}^-$ and $t_n = -n$ for each $n \in \mathbb{N}$), we have $R_\omega \subsetneq E_\omega^\circ$. □

We shall now give two general methods of constructing weights on $\mathbb{Z}$ by inductive processes. This will enable us to give several specific examples that illustrate certain possibilities. Throughout, we shall define a subadditive function η on $\mathbb{Z}$, with the understanding that the final weight is $\omega = \exp\eta$, as before. We shall define each η to be increasing and subadditive on $\mathbb{Z}^+$ (with $\eta(0) = 0$ throughout), and then extend η to all of the group $\mathbb{Z}$ by setting $\eta(-n) = \eta(n)$ $(n \in \mathbb{N})$, so that the final η is subadditive on $\mathbb{Z}$ and symmetric. We say that a function $\eta : \mathbb{Z}_k^+ \to \mathbb{R}^+$ is *subadditive to* k if $\eta(m+n) \leq \eta(m) + \eta(n)$ whenever $m, n, m+n \in \mathbb{Z}_k$. In fact it is more convenient to regard a function $\eta : \mathbb{Z}_k^+ \to \mathbb{R}^+$ as being defined on the interval $[0, k]$ of $\mathbb{R}^+$

so that we can use geometric language; of course we achieve this by joining the points $(j, \eta(j))$ to $(j+1, \eta(j+1))$ by a straight line.

The first remark is obvious.

PROPOSITION 9.9. *Let $\eta : [0, k] \to \mathbb{R}^+$ be convex, with $\eta(0) = 0$. Then η is subadditive to k.* □

For the first general construction, we again inductively define two strictly increasing sequences (m_k) and (n_k) in $\mathbb{N}$ such that

$$1 = m_1 < n_1 < \cdots < m_k < n_k < \cdots .$$

We shall impose the constraint that

$$n_k \geq 2m_k, \quad m_{k+1} = n_k + p_k \quad (k \in \mathbb{N}), \tag{9.6}$$

where $m_k \leq p_k \leq n_k$. We shall also arrange that η is increasing on $\mathbb{R}^+$ and constantly equal to 2^{k-1} on the interval $[m_k, n_k]$, so that $\eta(t) \to \infty$ as $t \to \infty$.

We set $\eta(0) = 0$ and $\eta(1) = 1$. Now assume inductively that m_k has been defined and that η is subadditive to m_k. Let $\widetilde{\eta}_k$ be a subadditive, increasing function on $[0, p_k]$, where $m_k \leq p_k \leq n_k$, such that

$$0 \leq \widetilde{\eta}_k(t) \leq \eta(t) \quad (0 \leq t \leq p_k) \quad \text{and} \quad \widetilde{\eta}_k(p_k) = \eta(p_k) = 2^{k-1} .$$

Define η on $[n_k, m_{k+1}]$ by setting

$$\eta(n_k + t) = \eta(n_k) + \widetilde{\eta}_k(t) = 2^{k-1} + \widetilde{\eta}_k(t) \quad (0 \leq t \leq p_k) .$$

A graph of the function η, regarded as a function on $\mathbb{R}^+$, is indicated in Figure 3, on page 121.

PROPOSITION 9.10. *The new function η is subadditive to m_{k+1}.*

PROOF. We must show that

$$\eta(m+n) \leq \eta(m) + \eta(n) \quad \text{whenever} \quad m, n, m+n \leq m_{k+1} ; \tag{9.7}$$

we may suppose that $m \leq n$. First suppose that $m_k \leq m + n \leq n_k$. If $m, n \leq m_k$, then there exists $r \in \mathbb{Z}^+_m$ with $r + n = m_k$, and then

$$\eta(m+n) = \eta(m_k) \leq \eta(r) + \eta(n) \leq \eta(m) + \eta(n) ,$$

and so (9.7) is clear. If $m \leq m_k \leq n$, then $\eta(m+n) = \eta(n)$, and so (9.7) follows. Second, suppose that $n_k \leq m + n \leq m_{k+1}$. We have $n \geq m_k$ because $n_k \geq 2m_k$. Set $m + n = n_k + r$, so that $r \leq m_k$. If $n \leq n_k$, then $m \geq r$, and so

$$\eta(m+n) - \eta(n) = 2^{k-1} + \widetilde{\eta}_k(r) - 2^{k-1} = \widetilde{\eta}_k(r) \leq \eta(m) ,$$

giving (9.7). If $n \geq n_k$, then $r = m + (n - n_k)$ and

$$\widetilde{\eta}(r) \leq \widetilde{\eta}_k(m) + \widetilde{\eta}_k(n - n_k) ,$$

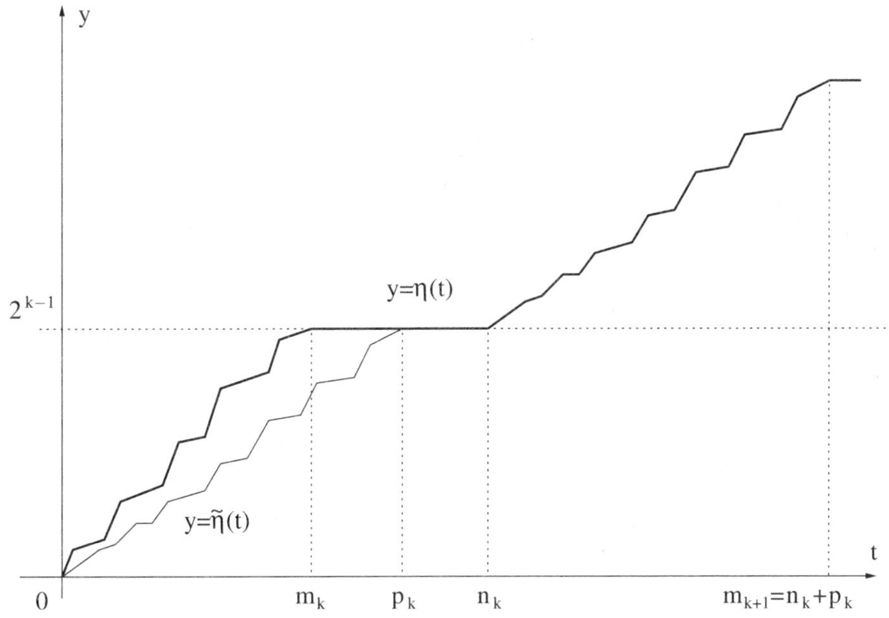

FIGURE 3

and so

$$\begin{aligned}\eta(m+n) &= 2^{k-1}+\widetilde{\eta}_k(r) \le 2^{k-1}+\widetilde{\eta}_k(m)+\widetilde{\eta}_k(n-n_k)\\ &= \widetilde{\eta}_k(m)+\eta(n) \le \eta(m)+\eta(n)\,,\end{aligned}$$

and again we have (9.7). This completes the proof of (9.7). □

As above, we extend η to $[0, n_{k+1}]$ by taking η to be constantly equal to 2^k on $[m_{k+1}, n_{k+1}]$. The next remark is clear.

PROPOSITION 9.11. *The new function η is subadditive to n_{k+1}.* □

We also note that we have the following inequality:

$$\Omega(m,n) \le \exp(\widetilde{\eta}_k(n)-\eta(n)) \quad (m_k \le m \le n_k,\ n \le m_k)\,. \tag{9.8}$$

EXAMPLE 9.12. *Let $\varphi : \mathbb{Z}^+ \to \mathbb{R}^+$ be an increasing function such that $\varphi(n) \to \infty$ as $n \to \infty$. Then there is a symmetric weight ω on $\mathbb{Z}$, increasing on $\mathbb{Z}^+$, such that $\omega(n) \to \infty$ and $\omega(n)/\varphi(n) \to 0$ as $n \to \infty$, such that the corresponding function Ω does not cluster on $\mathbb{Z}\times\mathbb{Z}$, such that ω is not almost invariant, and such that the algebra $\ell^1(\mathbb{Z},\omega)$ is not Arens regular.*

PROOF. Let $\psi : \mathbb{Z}^+ \to \mathbb{R}^+$ be such that $\exp\psi = \varphi$; we have

$$\psi(n) \to \infty \quad \text{as} \quad n \to \infty\,.$$

We apply the above algorithm, taking $p_k = m_k$ and $\widetilde{\eta} = \eta$; we choose n_k for $k \in \mathbb{N}$ to be such that $\psi(n_k) > k \cdot 2^k \quad (k \in \mathbb{N})$ (as well as requiring that $n_k \geq 2m_k \quad (k \in \mathbb{N})$). For each $n \in \mathbb{N}$ with $m_k \leq n \leq m_{k+1}$, we have

$$\eta(n)/\psi(n) \leq 2^k/\psi(n_k) < 1/k\,,$$

and so $\omega(n)/\varphi(n) \to 0$ as $n \to \infty$.

Certainly

$$\eta(n_k + 1) = \eta(n_k) + 1 \quad (k \in \mathbb{N})\,,$$

and so ω is not almost invariant.

Consider $\Omega(m_j, n_k)$. First, fix $k \in \mathbb{N}$. For each sufficiently large j, we have $\eta(m_j + n_k) = \eta(m_j)$, and so $\Omega(m_j, n_k) = 1/\omega(n_k)$. Second, fix $j \in \mathbb{N}$. For each sufficiently large k, we have

$$\eta(m_j + n_k) = \eta(m_j) + \eta(n_k)\,,$$

and so $\Omega(m_j, n_k) = 1$. Thus

$$\lim_k \lim_j \Omega(m_j, n_k) = 0, \quad \lim_j \lim_k \Omega(m_j, n_k) = 1\,,$$

and so Ω does not cluster on $\mathbb{Z} \times \mathbb{Z}$. By Theorem 8.11, A_ω is not Arens regular. □

Thus we can have symmetric weights ω on $\mathbb{Z}$ such that $\omega(n) \to \infty$ arbitrarily slowly and $\ell^1(\mathbb{Z}, \omega)$ is not Arens regular.

We wonder if the above example is strongly Arens irregular? If so, it would be the first such example that does not satisfy the conditions in either Corollary 8.16 or Theorem 8.18.

EXAMPLE 9.13. *Let $\varphi : \mathbb{Z}^+ \to \mathbb{R}^+$ be an increasing function such that $\varphi(n) \to \infty$ as $n \to \infty$. Then there is a symmetric weight ω on $\mathbb{Z}$, increasing on $\mathbb{Z}^+$, such that $\omega(n)/\varphi(n) \to 0$ as $n \to \infty$ and such that the algebra $\ell^1(\mathbb{Z}, \omega)$ is Arens regular.*

PROOF. We apply the above algorithm, taking $p_k = n_k$ and $\widetilde{\eta}_k(n)$ to be $n/2^{k-1} \quad (n \in \mathbb{Z}^+_{n_k})$ for each $k \in \mathbb{N}$; further, we choose n_k to be so large that $\widetilde{\eta}_k(n) \leq \eta(n)$ on $[0, n_k]$, so that $n_k > k \cdot 2^{k-1}$, and so that $\omega(n)/\varphi(n) < 1/k$ on $[0, n_k]$. Since $\lim_j \eta(j) = \infty$ and $\lim_j s(j) = 0$, it follows from Corollary 8.12 that $\ell^1(\mathbb{Z}, \omega)$ is Arens regular. □

Thus, given a weight ω on $\mathbb{Z}$ with $\operatorname{Lim} \omega(n) = \infty$, there is a symmetric weight $\overline{\omega}$ on $\mathbb{Z}$ such that $\ell^1(\mathbb{Z}, \omega) \subset \ell^1(\mathbb{Z}, \overline{\omega})$ and $\ell^1(\mathbb{Z}, \overline{\omega})$ is Arens regular.

EXAMPLE 9.14. *There is a symmetric weight ω on $\mathbb{Z}$, increasing on $\mathbb{Z}^+$, such that Ω 0-clusters on $\mathbb{N}\times\mathbb{N}$, but such that Ω does not 0-cluster strongly on $\mathbb{N}\times\mathbb{N}$. The Beurling algebra $\ell^1(\mathbb{Z},\omega)$ is Arens regular.*

PROOF. We apply the above algorithm, taking $p_k = n_k$ $(k \in \mathbb{N})$. For each $k \in \mathbb{N}$, we choose $\beta(k) \in \mathbb{N}_k$ such that $\beta(k) < k$ $(k \geq 2)$, and we now take $\widetilde{\eta}_k$ to be the function whose graph is formed from the two straight lines through the three points $(0,0)$, $(n_{\beta(k)}, 2^{\beta(k)-1})$, and $(n_k, 2^{k-1})$. We see that $\widetilde{\eta}_k$ is convex above $[0, n_k]$, so that $\widetilde{\eta}_k$ is subadditive to n_k, and that

$$\widetilde{\eta}_k(t) \leq \eta(t) \ \ (0 \leq t \leq n_k) .$$

We always obtain a function η on $\mathbb{Z}^+$ such that η is increasing, such that $\eta(m_k) = \eta(n_k) = 2^{k-1}$ $(k \in \mathbb{N})$, and such that η is subadditive on $\mathbb{Z}^+$.

We still have to make the choice of the numbers $\beta(k)$ for each $k \geq 2$. We make the choice to achieve the following. For each $r \in \mathbb{N}$, set

$$S_r = \{k \in \mathbb{N} : \beta(k) = r\} ,$$

so that $\{S_r : r \in \mathbb{N}\}$ is a family of pairwise disjoint subsets of $\mathbb{N}$. Our extra constraint is that each S_r is infinite and that $\bigcup\{S_r : r \in \mathbb{N}\} = \mathbb{N}$, so that $\{S_r : r \in \mathbb{N}\}$ is a partition of $\mathbb{N}$.

As usual, $\omega = \exp \eta$. We consider the function Ω on $\mathbb{N}\times\mathbb{N}$.

First, we see that Ω does not 0-cluster strongly on $\mathbb{N}\times\mathbb{N}$. To see this, note that $\Omega(n_k, n_{\beta(k)}) = 1$ for each $k \in \mathbb{N}$. Fix $r \in \mathbb{N}$. There are infinitely many values of $k \in \mathbb{N}$ with $\beta(k) = r$, and so $\limsup_m \Omega(m, n_r) = 1$. Thus

$$\limsup_n \limsup_m \Omega(m,n) = 1 ,$$

and so Ω does not 0-cluster strongly on $\mathbb{N}\times\mathbb{N}$.

Next we show that Ω 0-clusters on $\mathbb{N}\times\mathbb{N}$. Let us take sequences (u_i) and (v_j) in $\mathbb{N}$ such that

$$\alpha := \lim_j \lim_i \Omega(u_i, v_j)$$

exists; our aim is to show that $\alpha = 0$. We fix $n \in \mathbb{N}$, and look at the possible values of $\lim_m \Omega(m,n)$.

By passing to a subsequence of the sequence (u_i), we may suppose that (u_i) runs through values in the interval $[m_k, m_{k+1}]$, taking at most one value in each of these intervals; indeed, we may suppose that each u_i has the form $m_k + q_k$ for some $q_k \in \mathbb{N}$ with $m_k + q_k \leq m_{k+1}$ (the other cases are a trivial variation of this case). By (9.8), we see that

$$\lim_k \Omega(n_k + q_k, n) \leq \limsup_k \exp(\widetilde{\eta}_k(n) - \eta(n)) .$$

In the first subcase, suppose that infinitely many terms of the sequence (u_i) belong to the set $\{m_k + q_k : k \in S_r\}$ for some $r \in \mathbb{N}$; in fact, we may suppose that $u_i \in \{m_k + q_k : k \in S_r\}$ for each $i \in \mathbb{N}$; we may also suppose that $n > r$ because we shall eventually set $n = v_j$ and let $j \to \infty$, so that $v_j \to \infty$. We see by inspection of the graphs that $\lim_k \widetilde{\eta}_k(n) = \eta(r)$, and so

$$\lim_i \Omega(r_i, n) \leq \omega(r)/\omega(n) .$$

In the second subcase, suppose that only finitely many points of the sequence (u_i) belong to the set $\{m_k + q_k : k \in S_r\}$ for each $r \in \mathbb{N}$. Now we have $\widetilde{\eta}_k(n) = s_{\beta(k)} \cdot n$ for all $k \in \mathbb{N}$ and all sufficiently large $i \in \mathbb{N}$, where s_ℓ denotes the slope of the line from $(0,0)$ to $(n_{\beta(\ell)}, 2^{\beta(\ell)-1})$, and hence s_ℓ takes the value $2^{\beta(\ell)-1}/n_{\beta(\ell)}$. In this subcase, we have $\widetilde{\eta}_k(n) = 0$, and so $\lim_i \Omega(r_i, n) \leq 1/\omega(n)$. In both subcases, we have $\lim_j \lim_i \Omega(u_i, v_j) = 0$, and so $\alpha = 0$, as required.

This completes the proof of the example. □

EXAMPLE 9.15. *There is a symmetric weight ω on $\mathbb{Z}$, increasing on $\mathbb{Z}^+$, such that the Beurling algebra $\ell^1(\mathbb{Z}, \omega)$ is neither Arens regular nor strongly Arens irregular, such that $R_\omega = E_\omega^\circ$, such that*

$$\{0\} \neq R_\omega^{\square 2} \subset \mathfrak{Z}(B_\omega) ,$$

but such that $R_\omega^{\square 3} = \{0\}$.

PROOF. We apply the above algorithm, with the aditional constraints that

$$p_k = n_k, \quad n_k \geq 2m_k + k \cdot 2^k \quad (k \in \mathbb{N}) .$$

Note that, for each $k \in \mathbb{N}$, we have $2^k/n_{k+1} < 2^{k-1}/n_k$ because we required that $n_{k+1} > 2n_k$, and so the curve formed by joining each of the points $(n_{j-1}, \eta(n_{j-1}))$ to $(n_j, \eta(n_j))$ by a straight line for each $j \in \mathbb{N}_k$ is a convex curve, which we say defines the function $\widetilde{\eta}_k$ on $[0, n_k]$. By Proposition 9.9, the function $\widetilde{\eta}_k$ is subadditive to n_k. Thus $\widetilde{\eta}_k$ satisfies the conditions required in the above construction.

We note that, for each fixed $t \in \mathbb{N}$ and each sufficiently large $j \in \mathbb{N}$, we have

$$\widetilde{\eta}_k(n_j + t) = \eta(n_j) + 2^{j-1}t/(n_{j+1} - n_j) \leq \eta(n_j) + t/j .$$

As usual we set $A_\omega = \ell^1(\mathbb{Z}, \omega)$ and $B_\omega = A_\omega''$.

The values of the double sequence $(\Omega(m_j, n_k) : j, k \in \mathbb{N})$ have not changed from those that arose in Example 9.12 (where we took $\widetilde{\eta}_k = \eta$), and so it remains true that Ω does not cluster on $\mathbb{Z} \times \mathbb{Z}$. Thus, by Theorem 8.11, $\ell^1(\mathbb{Z}, \omega)$ is not Arens regular and $R_\omega^{\square 2} \neq \{0\}$.

We now consider the *claim* that

$$\lim_{t\to\infty}\liminf_{s\to\infty}\liminf_{r\to\infty}\alpha_{r,s,t}=\infty\,, \tag{9.9}$$

where

$$\alpha_{r,s,t}=\eta(r)+\eta(s)+\eta(t)-\eta(r+s+t)\quad (r,s,t\in\mathbb{N})\,.$$

First hold s and t fixed, and let $r\to\infty$. We consider $i\in\mathbb{N}$ such that $s+t\le m_i\le r$. First suppose that $r, r+s+t\in[m_i,n_i]$. Then $\alpha_{r,s,t}=\eta(s)+\eta(t)$, and so the claim is immediate. Second suppose that $r\in[m_i,n_i]$ and $r+s+t\in[n_i,m_{i+1}]$, say $r+s+t=n_i+r'$, where $r'\le s+t$. Then

$$\alpha_{r,s,t}=\eta(s)+\eta(t)-\widetilde{\eta}_k(r')\ge\eta(s)+\eta(t)-\widetilde{\eta}_k(s+t)\,.$$

Third suppose that $r, r+s+t\in[n_i,m_{i+1}]$, say $r=n_i+r'$. Then

$$\alpha_{r,s,t}=\widetilde{\eta}_k(r')+\eta(s)+\eta(t)-\widetilde{\eta}_k(r'+s+t)\ge\eta(s)+\eta(t)-\widetilde{\eta}_k(s+t)\,.$$

Finally suppose that $r\in[n_i,m_{i+1}]$ and $r+s+t\ge m_{i+1}$. Then there exist $s'\in\mathbb{N}_s$ and $t'\in\mathbb{N}_t$ with $r+s'+t'=m_{i+1}$ and such that $s',t'\to\infty$ as $s,t\to\infty$ and $\alpha_{r,s,t}=\alpha_{r,s',t'}$. Thus this case reduces to the previous one. Thus, we see that, to establish our claim, it suffices to show that

$$\lim_{t\to\infty}\liminf_{s\to\infty}\beta_{s,t}=\infty,\quad\text{where}\quad\beta_{s,t}=\eta(s)+\eta(t)-\widetilde{\eta}_k(s+t)\,. \tag{9.10}$$

We hold t fixed, and calculate $\lim_s\beta_{s,t}$. We consider $j\in\mathbb{N}$ such that $t\le m_j\le s$. Suppose that $s, s+t\in[m_j,n_j]$. Then

$$\eta(s)=\eta(s+t)\ge\widetilde{\eta}_k(s+t)\,,$$

and so $\beta_{s,t}\ge\eta(t)$, whence (9.10) is immediate. Suppose that we have $s\in[m_j,n_j]$ and $s+t\in[n_j,m_{j+1}]$. Then

$$\beta_{s,t}=\eta(n_j)+\eta(t)-\widetilde{\eta}_k(s+t)\ge\eta(n_j)+\eta(t))-\widetilde{\eta}_k(n_j+t)\,.$$

Suppose that $s, s+t\in[n_j,m_j]$, say $s=n_j+s'$. Then

$$\beta_{s,t}=\widetilde{\eta}_k(s')+\eta(t)-\widetilde{\eta}_k(s'+t)\ge\eta(n_j)+\eta(t)-\widetilde{\eta}_k(n_j+t)$$

because $\widetilde{\eta}_k$ is convex, $s'\le n_j$, and $\widetilde{\eta}_k(n_j)=\eta(n_j)$. In the above two cases, we have

$$\beta_{s,t}\ge\eta(t)-t/j$$

for $s\ge m_j$, and so $\liminf_s\beta_{s,t}\ge\eta(t)$. The final case is that in which $s\in[n_j,m_{j+1}-1]$ and $s+t\ge m_{j+1}$. Now we have

$$\beta_{s,t}=\eta(m_{j+1})-(m_{j+1}-s)2^{j-1}/n_j+\eta(t)-\eta(m_{j+1})\ge\eta(t)-2^{j-1}t/n_j\,,$$

and so again $\liminf_s\beta_{s,t}\ge\eta(t)$. Finally we see that (9.10) holds, and so we have established the claim (9.9).

Set $I = E_\omega^\circ$. In the terminology of Theorem 8.14, we have shown that the function Ω_3 0-clusters strongly on $\mathbb{Z}\times\mathbb{Z}\times\mathbb{Z}$, and so, by clause (ii) of that theorem, $I^{\square 3} = \{0\}$. Thus I is a nilpotent ideal of index 3, and hence $R_\omega = E_\omega^\circ$. It follows from Proposition 2.19 that $I^{\square 2} \subset \mathfrak{Z}(B_\omega)$ and hence that A_ω is not strongly Arens irregular.

This completes the proof of the example. □

We wonder if $R_\omega^{\square 2}$ is closed in the above example? It is possible that the centre $\mathfrak{Z}(B_\omega)$ is equal to $A_\omega \oplus R_\omega^{\square 2}$.

Our next example is based on a second general construction that we now explain.

Let (s_j) be a strictly decreasing sequence in $\mathbb{R}^+ \setminus \{0\}$, and let L_j be the half-line $y = s_j x \ (x \in \mathbb{R}^+)$, so that L_j has the slope s_j. We shall choose a strictly increasing sequence (m_j) in $\mathbb{N}$ by an inductive process. Once m_j is fixed, we shall specify (n_j) in $\mathbb{R}$ by requiring that

$$n_j s_{j+1} = m_j s_j \quad (j \in \mathbb{N})\,;$$

for convenience, we set $m_0 = n_0 = 0$. Note that

$$n_j - m_j = \left(\frac{s_j}{s_{j+1}} - 1\right) m_j \quad (j \in \mathbb{N}). \tag{9.11}$$

We then choose $m_{j+1} \in \mathbb{N}$ with $m_{j+1} > n_j$.

We now define $\eta : \mathbb{R}^+ \to \mathbb{R}^+$ by setting

$$\eta(t) = \begin{cases} s_j t & (n_{j-1} \leq t \leq m_j)\,, \\ s_j m_j & (m_j \leq t \leq n_j)\,, \end{cases}$$

for each $j \in \mathbb{Z}^+$. (The definitions of $\eta(m_k)$ are consistent.) Thus the graph of η lies on L_j over $[n_{j-1}, m_j]$ and is horizontal over $[m_j, n_j]$.

Clearly η is an increasing function on $\mathbb{R}^+$ with $\eta(0) = 0$. By choosing each m_j to be sufficiently large, we can ensure that we have $\eta(m_j) > \eta(m_{j-1}) + 1$, and so $\eta(t) \to \infty$ as $t \to \infty$.

A graph of the function η, regarded as a function on $\mathbb{R}^+$, is indicated in Figure 4 on page 127.

We *claim* that

$$\eta(s+t) \leq \eta(s) + \eta(t) \quad (s, t \in \mathbb{R}^+) \tag{9.12}$$

in each such case. To prove this, we may suppose that $s \leq t$. First, suppose that $n_{j-1} \leq t \leq m_j$ for some $j \in \mathbb{Z}^+$. Then $\eta(s) \geq s_j s$, $\eta(t) = s_j t$, and $\eta(s+t) \leq s_j(s+t)$, and so (9.12) holds. Second, suppose that $m_j \leq t \leq n_j$ for some $j \in \mathbb{N}$. Then $\eta(s) \geq s_{j+1} s$ and $\eta(t) \geq s_{j+1} t$; if also $s + t \geq n_j$, then $\eta(s+t) \geq s_{j+1}(s+t)$, and so

(9.12) holds in this case, and, if $s + t \leq n_j$, then $\eta(s+t) = \eta(t)$, and so (9.12) also holds in this case. Thus (9.12) holds in each case.

We now define $\eta(-t) = 0$ $(t \in \mathbb{R}^+)$, so that η is defined on $\mathbb{R}$. We claim that

$$\eta(s+t) \leq \eta(s) + \eta(t) \quad (s, t \in \mathbb{R}) . \tag{9.13}$$

Indeed (9.13) is clear if both $s, t \in \mathbb{R}^+$ and if both $s, t \in \mathbb{R}^-$. Now suppose that $s, t \in \mathbb{R}^+$. Then $\eta(-t) = 0$ and $\eta(s-t) \leq \eta(s)$ because η is increasing on $\mathbb{R}$. Thus (9.13) holds.

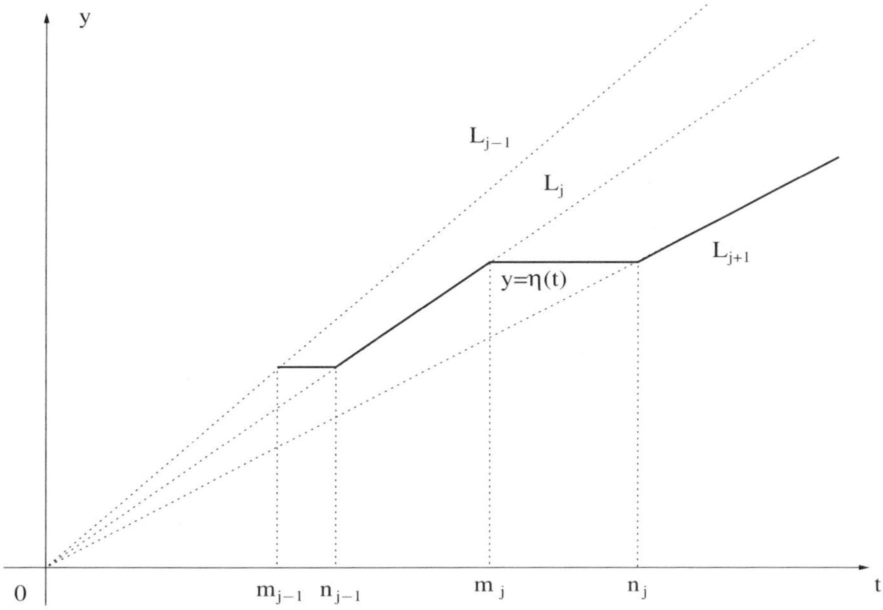

FIGURE 4

Consider the restriction of η to $\mathbb{Z}$: we obtain a subadditive function, and so $A_\omega = \ell^1(\mathbb{Z}, \omega)$ is a Beurling algebra on $\mathbb{Z}$ (where $\omega = \exp \eta$).

We note the following additional remark. Let η be as specified on $\mathbb{R}^+$, but now extend η to $\mathbb{R}$ by defining $\eta(-t) = \eta(t)$ $(t \in \mathbb{R}^+)$. It is immediate that η is also subadditive on $\mathbb{R}$, and $\omega = (\exp \eta) \mid \mathbb{Z}$ is a symmetric weight on $\mathbb{Z}$ such that there are sequences (s_j) and (t_j) in $\mathbb{N}$ with the properties that, for each $k_0 \in \mathbb{N}$, there exists $j_0 \in \mathbb{N}$ such that

$$\omega(s_j + k) = \omega(s_j), \quad \omega(t_j + k) = \omega(t_j)\omega(k) , \tag{9.14}$$

whenever $j \in \mathbb{N}$ with $j \geq j_0$ and $k \in \mathbb{Z}$ with $|k| \leq k_0$.

EXAMPLE 9.16. *There is a weight ω on $\mathbb{Z}$, increasing on $\mathbb{Z}^+$, such that $\omega(-n) = 1$ $(n \in \mathbb{N})$, but such that the Banach algebra $\ell^1(\omega)$ is neither Arens regular nor strongly Arens irregular. Further,*

$$\{0\} \neq R_\omega \subsetneq E_\omega^\circ .$$

PROOF. We choose a sequence (s_j) as above to satisfy the additional conditions that $s_j > 1/2$ $(j \in \mathbb{N})$ and $s_j \to 1/2$ as $j \to \infty$, and, further, such that

$$(9.15) \quad (s_j - 1/2)m_j \to \infty, \quad (s_j - s_{j+1})m_j/s_{j+1} \to \infty \quad \text{as } j \to \infty ;$$

this is clearly possible. It follows from (9.11) and (9.15) that

$$n_j - m_j \to \infty \quad \text{as} \quad j \to \infty .$$

The weight ω is defined as above. Since $\eta(-j) = 0$ $(j \in \mathbb{N})$, certainly both repeated limits of $(\Omega(-j,-k) : j,k \in \mathbb{N})$ are 1, and so A_ω is not Arens regular.

It remains to prove that A_ω is not strongly Arens irregular; for this, we again apply Corollary 8.10. As before, we define

$$\alpha_{j,k} = (\delta^1\eta)(m_k, j), \quad \beta_{j,k} = (\delta^1\eta)(m_k, -j) \quad (j,k \in \mathbb{N}),$$

so that, in fact, $\beta_{j,k} = \eta(m_k) - \eta(m_k - j)$ $(j,k \in \mathbb{N})$.

First, fix $j \in \mathbb{N}$. For k sufficiently large, we have $\eta(m_k + j) = \eta(m_k)$, and so $\lim_k \alpha_{j,k} = \eta(j)$. Hence $\lim_j \lim_k \alpha_{j,k} = \infty$.

Second, fix $k \in \mathbb{N}$. For each $j \in \mathbb{N}$, there exists $\ell \in \mathbb{Z}^+$ such that $m_\ell \leq j \leq m_{\ell+1}$. Now

$$\eta(j + m_k) - \eta(j) \leq s_\ell m_k ,$$

and so $\alpha_{j,k} \geq \eta(m_k) - s_\ell m_k$. Thus $\liminf_j \alpha_{j,k} \geq (s_k - 1/2)m_k$ because $s_\ell \to 1/2$ as $\ell \to \infty$. Hence $\lim_k \liminf_j \alpha_{j,k} = \infty$ by (9.15).

Third, fix $j \in \mathbb{N}$. For k sufficiently large, we have $m_k - j \geq n_{k-1}$, and so $\beta_{j,k} = s_k m_k - (s_k - j)m_j = j m_k$, whence $\lim_k \beta_{j,k} = \infty$. Hence $\lim_j \lim_k \beta_{j,k} = \infty$.

Finally, again fix $k \in \mathbb{N}$. Then $m_k - j < 0$ for j sufficiently large, and $\lim_j \beta_{j,k} = \eta(m_k)$. Hence $\lim_k \lim_j \beta_{j,k} = \infty$.

We have shown that all the required limits are ∞, and so A_ω is not strongly Arens irregular. By Corollary 8.10, $R_\omega \neq \{0\}$.

Again by Theorem 8.20 (with $S = \mathbb{Z}^-$ and $t_n = -n$ for each $n \in \mathbb{N}$), we have $R_\omega \subsetneq E_\omega^\circ$. We can also apply Theorem 8.27 with $S = \mathbb{Z}^-$ to see that $R_\omega \neq \{0\}$. □

The final example of this chapter is based on one suggested to us by Joel Feinstein, to whom we are grateful.

EXAMPLE 9.17. *There is a symmetric weight* ω *on* $\mathbb{Z}$ *such that* $\omega(0) = 1$ *and* $\omega(n) > 1$ $(n \in \mathbb{Z} \setminus \{0\})$, *such that* $\omega(n) = O(\log n)$ *as* $n \to \infty$, *and such that*

$$\liminf_{n\to\infty} \omega(n) \ < \infty\,,$$

but such that ω *is unbounded and not almost invariant. The algebra* $\ell^1(\omega)$ *is strongly Arens irregular and* $R_\omega \subsetneq E_\omega^\circ$.

PROOF. We again define an appropriate function η on $\mathbb{Z}$, and set $\omega = \exp\eta$.

Each $n \in \mathbb{Z}$ can be written (in many ways) in the form

$$n = \sum_{j=1}^{r} \varepsilon_j 2^{a_j}\,, \tag{9.16}$$

where $\varepsilon_j \in \{-1, 1\}$ and $a_j \in \mathbb{Z}^+$ for $j \in \mathbb{N}_r$, and where

$$a_1 \geq a_2 \geq \cdots \geq a_r\,.$$

For each $n \in \mathbb{Z} \setminus \{0\}$, we define $\eta(n)$ to be the minimum value of $r \in \mathbb{N}$ that can arise in equation (9.16); we also set $\eta(0) = 0$. We note that, when n has such a representation of minimum length, then necessarily

$$a_1 > a_2 > \cdots > a_r\,.$$

It is clear that $\eta(n) \in \mathbb{Z}^+$ $(n \in \mathbb{Z})$, that $\eta(n) = 0$ if and only if $n = 0$ (so that $\omega(n) > 1$ $(n \in \mathbb{Z}\setminus\{0\})$), and that $\eta(-n) = \eta(n)$ $(n \in \mathbb{Z})$ (so that ω is symmetric). It is also immediate that η is subadditive on $\mathbb{Z}$.

Clearly, $\eta(2^k) = 1$ and $\eta(2^k + 1) = 2$ for each $k \in \mathbb{N}$, and so

$$\liminf_{n\to\infty} \omega(n) = \mathrm{e}\,.$$

Thus ω is not almost invariant. Also it is immediate by induction on $k \in \mathbb{N}$ that $\eta(n) \leq k$ $(n \in \mathbb{N}_k)$, and so $\omega(n) = O(\log n)$ as $n \to \infty$.

Finally, we claim that η, and hence ω, is unbounded in $\mathbb{Z}$. To see this, consider numbers of the form

$$n_k = 2^{2k} + 2^{2k-2} + \cdots + 2^4 + 2^2 + 1 = (2^{2k+2} - 1)/3,$$

where $k \in \mathbb{N}$. We shall prove by induction on k that $\eta(n_k) = k + 1$. Certainly $\eta(n_k) \leq k + 1$. The result is immediate for $k = 1$; assume that the result holds for k. Suppose that

$$n_{k+1} = \sum_{j=1}^{r} \varepsilon_j 2^{a_j}, \tag{9.17}$$

where $\varepsilon_j \in \{-1, 1\}, a_j \in \mathbb{Z}^+$, and $a_1 > a_2 > \cdots > a_r \geq 0$. Since n_{k+1} is an odd number and $a_{r-1} \geq 1$, we have $a_r = 0$. Suppose that $\varepsilon_r = 1$. Then we subtract 1 from each side of (9.17), and note that necessarily $a_{r-1} \geq 2$ because the new right-hand side is divisible by 4. Then $n_k = \sum_{j=1}^{r-1} \varepsilon_j 2^{a_j - 2}$ with $a_{r-1} - 2 \geq 0$. Suppose that $\varepsilon_r = -1$. Then necessarily $a_{r-1} = 1$ and $\varepsilon_{r-1} = -1$. Further, we have

$$n_{k+1} = 1 + 4n_k = 4\left(\sum_{j=1}^{r-2} \varepsilon_j 2^{a_j - 2}\right) - 2 - 1,$$

and so $n_k = \sum_{j=1}^{r-2} \varepsilon_j 2^{a_j - 2} - 1$. In both these two cases, the inductive hypothesis shows that $r - 1 \geq k + 1$. Thus $\eta(n_{k+1}) = k + 2$, continuing the induction.

That A_ω is strongly Arens irregular follows from Corollary 8.16.

We *claim* that (8.11) of Theorem 8.20 is satisfied with $S = \mathbb{N}$ and with $t_n = 2^n$ $(n \in \mathbb{N})$. Indeed, for each fixed $s \in S$, we have

$$\eta(2^n + s) = 1 + \eta(s) = \eta(2^n) + \eta(s)$$

for n sufficiently large in $\mathbb{N}$, and so $\lim_n \Omega(s, t_n) = 1$, as required. Thus $R_\omega \subsetneq E_\omega^\circ$.

We have shown that our example has the required properties. $\square$

Unfortunately, we cannot see how to show that $R_\omega \neq \{0\}$ in the above example; the weight does not satisfy any of the sets of conditions for this that we have so far established.

It would be interesting to know if there is a weight ω on $\mathbb{Z}$ such that $(\ell^1(\omega)'', \Box)$ is a semisimple algebra; conceivably, the above example has this property.

CHAPTER 10

Beurling Algebras on $\mathbb{F}_2$

We now give some examples in the case where the underlying group is $\mathbb{F}_2$, the free group on two generators, and we first recall some standard facts about this group.

We denote the generators $\mathbb{F}_2$ by a and b. The *letters* of the group are a, a^{-1}, b, and b^{-1}, and a *word* is a finite formal product

$$s_1^{\varepsilon_1} \cdots s_m^{\varepsilon_m},$$

where $m \in \mathbb{N}$, $s_1, \ldots, s_m \in \{a, b\}$, and $\varepsilon_1, \ldots, \varepsilon_m \in \{1, -1\}$; the formal product with no factors is also a word, denoted by e. A word is *reduced* if it is e or if $\varepsilon_{k+1} = \varepsilon_k$ whenever $s_{k+1} = s_k$. Each word is equivalent to a reduced word in an obvious, unique way; the *length* of a word $w \in \mathbb{F}_2$ is the number of letters in the reduced word equivalent to w, and it is denoted by $|w|$ (with $|e| = 0$). The group $\mathbb{F}_2$ consists of all words; the product of two words is the reduced word equivalent to the word formed by juxtaposition of the two given words. Note that there are $4 \cdot 3^{m-1}$ words of length $m \in \mathbb{N}$ in $\mathbb{F}_2$.

Let (w_n) be a sequence in $\mathbb{F}_2$. Then clearly

$$\operatorname*{Lim}_n w_n = \infty \quad \text{if and only if} \quad \lim_{n\to\infty} |w_n| = \infty .$$

Now consider a general word w in $\mathbb{F}_2$, and suppose that w is reduced. Then w can be uniquely expressed in the form

$$w = (a^{j_1} b^{k_1})(a^{j_2} b^{k_2}) \cdots (a^{j_m} b^{k_m}), \tag{10.1}$$

where $m \in \mathbb{N}$, where $j_1, \ldots j_m, k_1, \ldots, k_m \in \mathbb{Z}$, and each of the integers $j_2, \ldots, , j_m, k_1, \ldots, k_{m-1}$ is non-zero. (The first and last terms could be just b^{k_1} and a^{j_m}, respectively.) We shall call this the *canonical representation* of w, and say that its *components* are $w_r = a^{j_r} b^{k_r}$ for $r \in \mathbb{N}_m$, so that $w = w_1 \cdots w_m$. We have

$$|w| = |w_1| + \cdots + |w_m| \quad \text{and} \quad |w_r| = |j_r| + |k_r| \quad (r \in \mathbb{N}_m) .$$

The words that have at most m components in their canonical representation are said to belong to the subset C_m of $\mathbb{F}_2$. Thus $C_0 = \{e\}$ and

$$C_1 = \{a^j b^k : j, k \in \mathbb{Z}\} .$$

Suppose that $u \in C_m$ and $v \in C_n$, where $m, n \in \mathbb{N}$. Then $uv \in C_{m+n}$.

EXAMPLE 10.1. *There is an unbounded weight ω on $\mathbb{F}_2$ such that ω is diagonally bounded on all of $\mathbb{F}_2$. Further, $\ell^1(\mathbb{F}_2, \omega)$ is strongly Arens irregular, but $\ell^1(\mathbb{F}_2, \omega)$ is not isomorphic to $\ell^1(\mathbb{F}_2)$.*

PROOF. In [J1, Proposition 2.8], Johnson constructs a function

$$\eta : \mathbb{F}_2 \to \mathbb{R}$$

such that $\eta(e_G) = 0$, such that

$$\left|(\delta^1\eta)(s,t)\right| \leq 6 \quad (s, t \in G),$$

such that

$$\eta(a^m) = \eta(b^n) = 0 \quad (m, n \in \mathbb{Z}),$$

and such that η is unbounded on G (indeed, $\eta((a^2b)^k) = 2k$ for each $k \in \mathbb{N}$). (Our function η is called α by Johnson.) Set

$$\omega(s) = \begin{cases} \exp(\eta(s) + 6) & (s \in G \setminus \{e_G\}), \\ 1 & (s = e_G). \end{cases}$$

Then, as in the remarks on page 94 after Theorem 7.44, ω is an unbounded weight which is diagonally bounded on all of $\mathbb{F}_2$.

By Theorem 8.15, $\ell^1(\mathbb{F}_2, \omega)$ is strongly Arens irregular.

Assume toward a contradiction that there is an isomorphism

$$\theta : \ell^1(\mathbb{F}_2, \omega) \to \ell^1(\mathbb{F}_2).$$

Then there is a function $\rho : \mathbb{F}_2 \to \mathbb{C}$ satisfying conditions (7.32), above, (with $G = \mathbb{F}_2$). By (7.32), we have $\rho(a^n) = n\rho(a)$ and $\rho(b^n) = n\rho(b)$ for $n \in \mathbb{Z}$, and so it follows from (7.32) that $\rho(a) = \rho(b) = 0$, and hence that $\rho = 0$. But now we conclude from (7.32) that η is bounded on $\mathbb{F}_2$, a contradiction. Thus $\ell^1(\mathbb{F}_2, \omega)$ is not isomorphic to $\ell^1(\mathbb{F}_2)$. □

The above example was known to at least B. E. Johnson and M. C. White. More general examples than the above can be constructed by using results from the theory of bounded group cohomology, as expounded by Grigorchuk in [Gri].

EXAMPLE 10.2. *There is a weight on $\mathbb{F}_2$ which is almost left-invariant, but not almost invariant.*

PROOF. Let S be the set of (reduced) words in $\mathbb{F}_2$ which end in a, and define η in terms of S as in (7.9), so that $\exp \eta$ is a weight on $\mathbb{F}_2$.

Let $s \in \mathbb{F}_2$. For each $t \in \mathbb{F}_2$ with $|t| > |s| + 1$, the last letter of the reduced form of st is the same as the last letter of t, and so

$\eta(st) = \eta(t)$. Thus $\operatorname{Lim}_{t\to\infty}(\eta(st) - \eta(t)) = 0$, and so ω is almost left-invariant. However,

$$\eta(b^n a) - \eta(b^n) = 2 - 1 = 1 \quad (n \in \mathbb{N}),$$

and so it is not true that $\operatorname{Lim}_{t\to\infty}(\eta(ta) - \eta(t)) = 0$. This shows that ω is not almost invariant. □

The above weight is obviously equivalent to an invariant weight; we cannot see an almost left-invariant weight which is not equivalent to an almost invariant weight.

We now approach our main example on the group $\mathbb{F}_2$. We shall first present some preliminary lemmas.

We start with a function φ which is continuous on $\mathbb{R}^+$, twice-differentiable on $\mathbb{R}^+ \setminus \{0\}$, unbounded, and such that $\varphi(0) = 0$ and $\varphi'(t) > 0$ and $\varphi''(t) < 0$ for each $t > 0$. Set $\psi = \varphi/Z$, so that $\psi(t) = \phi(t)/t$ for $t > 0$. It is clear that φ is increasing and convex on $\mathbb{R}^+$ and that $\varphi(t) \to \infty$, and $\varphi(t+1) - \varphi(t) \to 0$ as $t \to \infty$. Also ψ is decreasing on $\mathbb{R}^+ \setminus \{0\}$ and $\psi(t) \to 0$ as $t \to \infty$. For example, set

$$\varphi_\alpha(t) = t^\alpha \quad (t \geq 0),$$

where $0 < \alpha < 1$. Then φ satisfies all the required conditions. We shall also require that

$$(10.2) \qquad t \mapsto \left(1 + \frac{j}{t}\right)\varphi(t) \quad \text{is increasing on} \quad [j, \infty)$$

for each $j \in \mathbb{N}$. For example, the above function φ_α satisfies this extra condition if and only if $\alpha \geq 1/2$.

We define

$$\alpha_j = \varphi(|j|) \quad (j \in \mathbb{Z}).$$

We see immediately that $\alpha_0 = 0$, that

$$\alpha_{j+k} \leq \alpha_j + \alpha_k \quad (j, k \in \mathbb{Z}),$$

and that $\alpha_{-j} = \alpha_j$ $(j \in \mathbb{Z})$. Set $\alpha = (\alpha_j : j \in \mathbb{Z})$. Then $\exp \alpha$ is a weight on $\mathbb{Z}$.

We note that

$$(10.3) \qquad \lim_m \lim_n (\alpha_m + \alpha_n - \alpha_{m+n}) = \infty.$$

Indeed, for $n \geq m$, we have $\alpha_n \geq (n-m)\alpha_{m+n}/n$ because φ is convex, and so $\alpha_n - \alpha_{m+n} \geq -m\alpha_{m+n}/n \to 0$ as $n \to \infty$. Now (10.3) follows because $\lim_m \alpha_m = \infty$.

We shall now define a function

$$(j,k) \mapsto \alpha(j,k), \quad \mathbb{Z} \times \mathbb{Z} \to \mathbb{R}^+.$$

To do this, we divide the space $\mathbb{Z} \times \mathbb{Z}$ into two parts by drawing the lines $k = \pm j$, and using different formulae on the two parts.

First, suppose that $|j| \geq |k|$. Then

$$\alpha(j, k) = \alpha_j + \alpha_k \quad (|j| \geq |k|) \,. \tag{10.4}$$

Second, suppose that $|j| \leq |k|$, and first set $\alpha(0, k) = \alpha_k$ $(k \in \mathbb{Z})$. Then extend $\alpha(j, k)$ to be linear in j on the intervals $[0, |k|]$ and $[-|k|, 0]$ for each $k \in \mathbb{Z}$. Thus

$$\alpha(j, k) = \left(1 + \frac{|j|}{|k|}\right) \alpha_k \quad (0 < |j| \leq |k|) \,. \tag{10.5}$$

Suppose that $|j_n| + |k_n| \to \infty$ as $n \to \infty$. Then clearly we have $\alpha(j_n, k_n) \to \infty$ as $n \to \infty$.

LEMMA 10.3. (i) *For each* $k \in \mathbb{Z}$, *the function* $j \mapsto \alpha(j, k)$ *is increasing on* $\mathbb{Z}^+$.

(ii) *For each* $j \in \mathbb{Z}$, *the function* $k \mapsto \alpha(j, k)$ *is increasing on* $\mathbb{Z}^+$.

PROOF. (i) This is immediate.

(ii) It suffices to show that $\alpha(j, k_1) \leq \alpha(j, k_2)$ whenever we have $k_2 \geq k_2 \geq j > 0$. We require that

$$\left(1 + \frac{j}{k_1}\right) \alpha_{k_1} \leq \left(1 + \frac{j}{k_2}\right) \alpha_{k_2} \,.$$

But this follows immediately from (10.2). □

LEMMA 10.4. (i) $\alpha(j, k) \leq \alpha_j + \alpha_k$ $(j, k \in \mathbb{Z})$.

(ii) $\alpha(j_1 + j_2, k) \leq \alpha(j_1, 0) + \alpha(j_2, k)$ $(j_1, j_2, k \in \mathbb{Z})$.

(iii) $\alpha(j, k_1 + k_2) \leq \alpha(j, k_1) + \alpha(0, k_2)$ $(j, k_1, k_2 \in \mathbb{Z})$.

PROOF. (i) This is the immediate from (10.4) in the case where $|j| \geq |k|$. Now suppose that $|j| \leq |k|$. Then we require that

$$(1 + |j| \,/\, |k|)\alpha_k \leq \alpha_j + \alpha_k \,,$$

and hence that $(|j| \,/\, |k|)\alpha_k \leq \alpha_j$. This holds because (α_n/n) is a decreasing sequence.

(ii) Suppose that $j_2 \geq k \geq 0$ and $j_1 \geq 0$. Then we require that

$$\alpha_{j_1 + j_2} + \alpha_k \leq \alpha_{j_1} + \alpha_{j_2} + \alpha_k \,,$$

which is true.

Now suppose that $0 \leq j_2 \leq k$ and $j_1 \geq 0$. If $j_1 + j_2 \leq k$, then we require that

$$\left(1 + \frac{j_1 + j_2}{k}\right) \alpha_k \leq \alpha_{j_1} + \left(1 + \frac{j_2}{k}\right) \alpha_{k_2} \,,$$

and this is true because $(j_1/k)\alpha_k \leq \alpha_{j_1}$. If $j_1 + j_2 \geq k$, then we require that

$$\alpha_{j_1+j_2} + \alpha_k \leq \alpha_{j_1} + \left(1 + \frac{j_2}{k}\right) \alpha_k \,,$$

and hence that

$$\alpha_{j_1} + \left(\frac{j_2}{k}\right) \alpha_k - \alpha_{j_1+j_2} \geq 0 \,.$$

The left-hand side of the above inequality is a decreasing function of j_1, and so we may suppose that $j_1 + j_2 = k$. Again we see that we require that $\alpha_{j_1} \geq (j_1/k)\alpha_k$, which is true.

Finally we require that $\alpha(|j_1 - j_2|, k) \leq \alpha(j_1, 0) + \alpha(j_2, k)$ whenever $j_1, j_2, k \geq 0$. This follows from Lemma 10.3(i).

(iii) Suppose that $k_1 \geq j \geq 0$ and $k_2 \geq 0$. Then we require that

$$\left(1 + \frac{j}{k_1 + k_2}\right) \alpha_{k_1+k_2} \leq \left(1 + \frac{j}{k_1}\right) \alpha_{k_1} + \alpha_{k_2} \,,$$

and this is true because

$$\alpha_{k_1+k_2} \leq \alpha_{k_1} + \alpha_{k_2} \quad \text{and} \quad \alpha_{k_1+k_2}/(k_1 + k_2) \leq \alpha_{k_1}/k_1 \,.$$

Now suppose that $0 \leq k_1 \leq j$ and $k_2 \geq 0$. If $k_1 + k_2 \geq j$, then we require that

$$\left(1 + \frac{j}{k_1 + k_2}\right) \alpha_{k_1+k_2} \leq \alpha_j + \alpha_{k_1} + \alpha_{k_2} \,,$$

which is true. If $k_1 + k_2 \leq j$, then we require that

$$\alpha_{k_1+k_2} + \alpha_j \leq \alpha_j + \alpha_{k_1} + \alpha_{k_2} \,,$$

which is also true.

Finally, we require that $\alpha(j, |k_1 - k_2|) \leq \alpha(j, k_1) + \alpha(0, k_2)$ whenever $j, k_1, k_2 \geq 0$. This follows from Lemma 10.3(ii). □

We now define η on the subset C_1 of $\mathbb{F}_2$ by the formula

$$\eta(a^j b^k) = \alpha(j, k) \quad (j, k \in \mathbb{Z}) \,. \tag{10.6}$$

Note that $\eta(e) = \alpha(0, 0) = 0$.

LEMMA 10.5. *Suppose that u, v, and uv belong to C_1. Then*

$$\eta(uv) \leq \eta(u) + \eta(v) \,.$$

PROOF. The only cases which arise are: (i) $u = a^j$ and $v = b^k$; (ii) $u = a^{j_1}$ and $v = a^{j_2}b^k$; (iii) $u = a^jb^{k_1}$ and $v = b^{k_2}$ for some $j, k, j_1, j_2, k_1, k_2 \in \mathbb{Z}$. The result in each of these three cases is immediate from Lemma 10.4. □

We now define $\eta(w)$ for an arbitrary element w of $\mathbb{F}_2$. Let w have the canonical representation $w = w_1 \cdots w_m$, as described above. Then $\eta(w_r)$ has already been defined for each $r \in \mathbb{N}_m$.

DEFINITION 10.6. *Let $w = w_1 \cdots w_m$ be the canonical representation of an element $w \in \mathbb{F}_2$. Then*

$$\eta(w) = \left(\eta(w_1)^2 + \cdots + \eta(w_m)^2\right)^{1/2}.$$

Let (s_m) be a sequence in $\mathbb{F}_2$ such that $|s_m| \to \infty$ as $m \to \infty$. Then $\eta(s_m) \to \infty$ as $m \to \infty$.

LEMMA 10.7. *The function η is subadditive on $\mathbb{F}_2$.*

PROOF. Let $u, v \in \mathbb{F}_2$, with canonical representations $u = u_1 \cdots u_m$ and $v = v_1 \cdots v_n$, say.

First suppose that the canonical representation of uv is

$$u_1 \cdots u_m v_1 \cdots v_n\,,$$

with no cancellation, so that uv has $m + n$ components. Then we require that

$$\text{(10.7)} \quad \left\{ \begin{array}{l} (x_1^2 + \cdots + x_m^2 + y_1^2 + \cdots + y_n^2)^{1/2} \\ \qquad \leq (x_1^2 + \cdots + x_m^2)^{1/2} + (y_1^2 + \cdots + y_n^2)^{1/2}, \end{array} \right.$$

where $x_r = \eta(u_r)$ and $y_s = \eta(v_s)$. But (10.7) is certainly true.

Second, suppose that there is some cancellation in the canonical representation of uv. For example, suppose that

$$u = u'a^{j_1} \quad \text{and} \quad v = a^{j_2}b^kv'$$

for some $j_1, j_2, k \in \mathbb{Z}$ and some $u', v' \in \mathbb{F}_2$. Suppose further that $j_1 + j_2 \neq 0$, so that the canonical representation of uv is $u'(a^{j_1+j_2}b^k)v'$. Then we require that

(10.8)

$$(x + \alpha(j_1 + j_2, k)^2 + y)^{1/2} \leq (x + \alpha(j_1, 0)^2)^{1/2} + (\alpha(j_2, k)^2 + y)^{1/2}$$

for certain $x, y \geq 0$. But this follows from Lemma 10.4(ii). In the case where $j_1 + j_2 = 0$, there is further cancellation, but a similar argument applies. An extreme case where the number of components in uv is strictly less than $m + n$ is covered in Lemma 10.5.

There may be cancellation because the juxtaposition of u and v does not give a reduced word. For example, suppose that we have $u = a^{j+r}b^s$ and $v = b^{-s}a^{-r}b^k$ for some $j, k, r, s \in \mathbb{N}$. Now $uv = a^j b^k$, and so the inequality requires that

$$\alpha(j,k) \leq \alpha(j+r,s) + (\alpha(0,s)^2 + \alpha(r,k)^2)^{1/2}.$$

The right-hand side of this inequality attains its minimum (as a function of r and s) at $r = s = 0$, and so it suffices to note that we have $\alpha(j,k) \leq \alpha_j + \alpha_k$. □

At this stage, we have established that $\omega = \exp \eta$ is a weight on the group $\mathbb{F}_2$. We note that ω is not symmetric on $\mathbb{F}_2$. For example, suppose that $j \geq k > 0$. Then $\eta(a^j b^k) = \alpha_j + \alpha_k$, but

$$\eta((a^j b^k)^{-1}) = \eta(b^{-k}a^{-j}) = (\alpha_j^2 + \alpha_k^2)^{1/2} \neq \alpha_j + \alpha_k.$$

We now set $A_\omega = \ell^1(\mathbb{F}_2, \omega)$, etc., as before, so that $A_\omega = E'_\omega$.

LEMMA 10.8. *Let* $\Phi, \Psi \in E^\circ_\omega$. *Then* $\Phi \,\Box\, \Psi = 0$.

PROOF. By a remark after Theorem 7.8, it suffices to show that

$$\lim_m \liminf_n (\eta(s_m) + \eta(t_n) - \eta(s_m t_n)) = \infty \tag{10.9}$$

whenever (s_m) and (t_n) are sequences in $\mathbb{F}_2$ such that $|s_m| \to \infty$ and $|t_n| \to \infty$. Take such sequences (s_m) and (t_n).

First, as in the proof of Lemma 10.7, suppose that the canonical representation of each product $s_m t_n$ is formed by the juxtaposition of the canonical representations of s_m and t_n. Then we must consider

$$\lim_m \liminf_n \left(f(m)^{1/2} + g(n)^{1/2} - (f(m) + g(n))^{1/2}\right), \tag{10.10}$$

where $f, g : \mathbb{N} \to \mathbb{R}^+$ are functions with the properties that $f(m) \to \infty$ as $m \to \infty$ and $g(n) \to \infty$ as $n \to \infty$. Fix $m \in \mathbb{N}$. Then

$$f(m)^{1/2} + g(n)^{1/2} - (f(m) + g(n))^{1/2} = g(n)^{1/2}\left(1 + t - (1+t)^{1/2}\right),$$

where $t = f(m)^{1/2}/g(n)^{1/2}$. Since

$$1 + t - (1+t)^{1/2} \geq t/2 \quad (t \geq 0),$$

the right-hand side of the above inequality is at least $f(m)^{1/2}/2$, and so the limit in (10.10) is ∞, as required.

Again, we must allow for some cancellation when the canonical representation of s_m, t_n is formed.

For example, suppose that $s_m = a^m$ and $t_n = b^n$. Let $m, n \in \mathbb{N}$ with $m \geq n$. Then $\eta(a^m b^n) = (1 + m/n)\alpha_n$, and so, for each $m \in \mathbb{N}$, we have

$$\begin{aligned}\lim_n(\eta(a^m) + \eta(b^n) - \eta(a^m b^n)) &= \lim_n(\alpha_m + \alpha_n - (1 + m/n)\alpha_n) \\ &= \lim_n(\alpha_m - \alpha_n/n) = \alpha_m\,.\end{aligned}$$

Thus the limit in (10.9) is $\lim_m \alpha_m$, which is ∞, as required. Again, suppose that $s_m = a^m$ and $t_n = a^n$ (or $s_m = b^m$ and $t_n = b^n$). Then the limit in (10.9) is ∞ by (10.3).

Now suppose that there is some intermediate level of cancellation when the canonical representation of $s_m t_n$ is formed; we must consider analogues of equation (10.8). We then see that the result follows by combinations of the two arguments already given. □

It follows that

$$\mathfrak{Z}_t^{(1)}(B_\omega) = A_\omega \oplus \{\Phi \in E_\omega^\circ : \Phi \diamond \Psi = 0 \ \ (\Psi \in E_\omega^\circ)\}$$

and

$$\mathfrak{Z}_t^{(2)}(B_\omega) = A_\omega \oplus \{\Phi \in E_\omega^\circ : \Psi \diamond \Phi = 0 \ \ (\Psi \in E_\omega^\circ)\}\,.$$

LEMMA 10.9. *Let $\Phi_1, \Phi_2, \Phi_3 \in E_\omega^\circ$. Then $\Phi_1 \diamond \Phi_2 \diamond \Phi_3 = 0$.*

PROOF. By Theorem 8.14(ii), it suffices to show that

$$\text{(10.11)} \quad \operatorname*{Lim}_k \operatorname*{Lim\,inf}_j \operatorname*{Lim\,inf}_i (\eta(r_i) + \eta(s_j) + \eta(t_k) - \eta(r_i s_j t_k)) = \infty$$

whenever (r_i), (s_j), and (t_k) are sequences in $\mathbb{F}_2$ such that $|r_i| \to \infty$, $|s_j| \to \infty$, and $|t_k| \to \infty$.

In the case where there is no cancellation when the canonical representation of $r_i s_j t_k$ is formed from the juxtaposition of the canonical representations of r_i, s_j, and t_k, the argument for (10.11) is the same as that for (10.10).

An extreme case in the opposite direction is that in which we have $r_i s_j t_k \in C_1$. For example, we could have $r_i = a^i$, $s_j = a^j$, and $t_k = b^k$. We then require the limit

$$\operatorname*{Lim}_k \operatorname*{Lim\,inf}_j \operatorname*{Lim\,inf}_i (\alpha_i + \alpha_j + \alpha_k - \alpha(i + j, k))\,.$$

By Lemma 10.4(i), $\alpha(i + j, k) \leq \alpha_{i+j} + \alpha_k$, and so this limit is at least

$$\lim_j \lim_i (\alpha_i + \alpha_j - \alpha_{i+j})\,;$$

by (10.3), this limit is ∞.

All other cases follow by a combination of the above arguments. □

LEMMA 10.10. *There exist* $\Phi_0, \Psi_0 \in E_\omega^\circ$ *such that* $\langle \Phi_0 \diamond \Psi_0,\, \omega\rangle = 1$.

PROOF. Consider

$$\lim_n \lim_m (\eta(a^m) + \eta(b^n) - \eta(a^m b^n))\,.$$

For each $m \geq n$, we have $\eta(a^m b^n) = \alpha_m + \alpha_n = \eta(a^m) + \eta(b^n)$, and so this repeated limit is 0. Let Φ_0 and Ψ_0 be weak-$*$ accumulation points of $(\widetilde{\delta}_{a^m})$ and $(\widetilde{\delta}_{b^n})$, respectively. Then

$$\langle \Phi_0 \diamond \Psi_0, \omega\rangle = \lim_n \lim_m \Omega(a^m, b^n) = \exp 0 = 1\,,$$

as required. □

We now regard Φ_0 and Ψ_0 as fixed elements of $\mathcal{B}_\omega$.

LEMMA 10.11. (i) $\Psi \diamond \Phi_0 = 0 \;\; (\Psi \in E_\omega^\circ)$, *and so* $\Phi_0 \in \mathfrak{Z}_t^{(2)}(B_\omega)$.
(ii) $\Psi_0 \diamond \Psi = 0 \;\; (\Psi \in E_\omega^\circ)$, *and so* $\Psi_0 \in \mathfrak{Z}_t^{(1)}(B_\omega)$.
(iii) $\Phi_0 \notin \mathfrak{Z}_t^{(1)}(B_\omega)$ *and* $\Psi_0 \notin \mathfrak{Z}_t^{(2)}(B_\omega)$.

PROOF. (i) Let (s_m) be a sequence in $\mathbb{F}_2$ with $|s_m| \to \infty$. By a small variation of earlier arguments, we see that

$$\lim_n \lim_m (\eta(s_m) + \eta(a^n) - \eta(s_m a^n)) = \infty\,,$$

and this is sufficient to establish that $\Psi \diamond \Phi_0 = 0 \;\; (\Psi \in E_\omega^\circ)$. We certainly have $\Psi \,\Box\, \Phi_0 = 0 \;\; (\Psi \in E_\omega^\circ)$, and so $\Phi_0 \in \mathfrak{Z}_t^{(2)}(B_\omega)$.
(ii) This is similar.
(iii) We have $\Phi_0 \,\Box\, \Psi_0 = 0$, but $\Phi_0 \diamond \Psi_0 \neq 0$. □

It follows that $\Phi_0^{\diamond 2} = \Psi_0^{\diamond 2} = 0$, and so $(\Phi_0 + \Psi_0)^{\diamond 2} = \Phi_0 \diamond \Psi_0 \neq 0$.

We can now summarize the properties of this example that we have established.

THEOREM 10.12. *There is a weight on* ω *on the group* $\mathbb{F}_2$ *such that the following properties hold:*
(i) $(E_\omega^\circ)^{\Box 2} = \{0\}$;
(ii) $(E_\omega^\circ)^{\diamond 3} = \{0\}$;
(iii) *there exist* $\Phi_0, \Psi_0 \in E_\omega^\circ$ *such that* $\Phi_0 \diamond \Psi_0 \neq 0$;
(iv) $\Phi_0 \in \mathfrak{Z}_t^{(2)}(B_\omega) \setminus \mathfrak{Z}_t^{(1)}(B_\omega)$ *and* $\Psi_0 \in \mathfrak{Z}_t^{(1)}(B_\omega) \setminus \mathfrak{Z}_t^{(2)}(B_\omega)$, *and so*

$$\mathfrak{Z}_t^{(1)}(B_\omega) \neq \mathfrak{Z}_t^{(2)}(B_\omega)\,;$$

(v) $\operatorname{rad}(B_\omega,\, \Box\,) = \operatorname{rad}(B_\omega,\, \diamond\,) = E_\omega^\circ$. □

Set $R_\omega^\diamond = \operatorname{rad}(B_\omega, \diamond)$, as before. We have shown that $(R_\omega^\diamond)^2 \neq \{0\}$; it is not clear whether or not $(R_\omega^\diamond)^2$ is closed in B_ω.

We see that, in the present example, the two radicals of $(B_\omega, \Box)$ and $(B_\omega, \diamond)$ are the same subset of B_ω (but they have different algebraic properties). We have not been able to construct an example in which they are distinct sets.

EXAMPLE 10.13. Let $A = \ell^1(\mathbb{F}_2, \omega)$, with ω as above. Then we certainly have $\mathfrak{Z}_t^{(1)}(A'') \neq \mathfrak{Z}_t^{(2)}(A'')$. For $f \in A$, set

$$\overline{f}(s) = \overline{f(s)} \quad (s \in \mathbb{F}_2).$$

Then the map $f \mapsto \overline{f}$ is a linear involution on A such that

$$\overline{f \star g} = \overline{f} \star \overline{g} \quad (f, g \in A)$$

As in Example 4.4, we can construct a Banach $*$-algebra $C = A \oplus A^{\mathrm{op}}$ such that $\mathfrak{Z}_t^{(1)}(C'') \neq \mathfrak{Z}_t^{(2)}(C'')$. In this case, we do have

$$\operatorname{rad}(C'', \Box) = \operatorname{rad}(C'', \diamond). \qquad \Box$$

CHAPTER 11

Topological Centres of Duals of Introverted Subspaces

In this chapter, we shall establish two results on the topological centre of left-introverted subspaces.

First, let G be a group, and let X be a left-introverted subspace of $A'_\omega = \ell^\infty(G, 1/\omega)$. Then, in the case where Ω clusters on $G \times G$, we have $\mathfrak{Z}_t(X') = X'$ whenever we have $X \subset WAP(G, 1/\omega)$. Second, let G be a locally compact group, and let $\mathcal{X}_\omega = LUC(G, 1/\omega)$, as before. Then, in Theorem 11.9, we shall identify $\mathfrak{Z}_t(\mathcal{X}'_\omega)$ as $M(G, \omega)$ under certain conditions on the weight function ω.

Let ω be a weight on a group G. We continue to use the notations A_ω, A'_ω, B_ω, and E_ω that were introduced in (8.1). Recall also that $\ell_t\lambda = \lambda \cdot \delta_t$ and $r_t\lambda = \delta_t \cdot \lambda$ for each $t \in G$ and $\lambda \in A'_\omega$. We also define

$$\left.\begin{array}{rcccl} \Phi_\ell(\lambda)(t) & = & \langle \Phi, \ell_t\lambda \rangle & = & (\Phi \cdot \lambda)(t)\,, \\ \Phi_r(\lambda)(t) & = & \langle \Phi, r_t\lambda \rangle & = & (\lambda \cdot \Phi)(t)\,, \end{array}\right\} \quad (t \in G,\ \lambda \in A'_\omega,\ \Phi \in B_\omega).$$

A left-introverted subspace X of A'_ω was defined in Definition 5.1. A $\|\cdot\|$-closed subspace X is an A_ω-submodule of A'_ω if and only if it is *translation-invariant*, in the sense that $\ell_t\lambda, r_t\lambda \in X$ whenever $t \in G$ and $\lambda \in X$. In this case, X is left-introverted if and only if $\Phi_\ell(\lambda) \in X$ whenever $\lambda \in X$ and $\Phi \in X'$. To show that a translation-invariant subspace X of A'_ω is left-introverted, it suffices to show that

$$\Phi \cdot \lambda \in X \quad (\lambda \in X,\ \Phi \in S_\omega) \tag{11.1}$$

because $B_\omega = \operatorname{lin} S_\omega$. Suppose that X is a left-introverted subspace of A'_ω. Then it follows from Theorem 5.4 that X' is a Banach algebra for the product $\Box$, defined for $\Phi, \Psi \in X'$ by the equation

$$\langle \Phi \,\Box\, \Psi, \lambda \rangle = \langle \Phi, \Psi \cdot \lambda \rangle \quad (\lambda \in X)\,,$$

given as (5.1).

The topologies of X and X' are again taken to be the weak-$*$ topologies $\sigma(X, A_\omega)$ and $\sigma(X', X)$, respectively, unless stated otherwise.

Let $\lambda \in A'_\omega$. We denote by $\overline{\langle K_\lambda \rangle}$ the weak-$*$ closed, convex hull in A'_ω of the set

$$K_\lambda = \left\{ \widetilde{\delta}_s \cdot \lambda : s \in G \right\}. \tag{11.2}$$

The first lemma was proved in [GrL, Lemma 2] in the case where $\omega = 1$; the result is similar to Proposition 5.3. We use the following notation. Let $\gamma = \sum_{i=1}^n c_i \widetilde{\delta}_{s_i} \cdot \lambda$ be an element of $\langle K_\lambda \rangle$, so that $c_1, \dots, c_n \geq 0$ and $\sum_{i=1}^n c_i = 1$. Then the corresponding element $\widetilde{\gamma}$ of A_ω is

$$\widetilde{\gamma} = \sum_{i=1}^n c_i \, \widetilde{\delta}_{s_i} .$$

Notice that, for each $t \in G$, we have

$$\langle \widetilde{\gamma}, \lambda \cdot \delta_t \rangle = \sum_{i=1}^n c_i \langle \widetilde{\delta}_{s_i}, \lambda \cdot \delta_t \rangle = \gamma(t) .$$

LEMMA 11.1. *Let G be a group, and let ω be a weight on G. A $\|\cdot\|$-closed, left-translation invariant subspace X of A'_ω is left-introverted if and only if $\overline{\langle K_\lambda \rangle} \subset X$ for each $\lambda \in X$.*

PROOF. Suppose that X is left-introverted, and take $\lambda \in X$ and $\gamma \in \overline{\langle K_\lambda \rangle}$. Then there is a net (γ_α) in $\langle K_\lambda \rangle$ such that $\gamma_\alpha \to \gamma$ in $(A'_\omega, \sigma(A'_\omega, A_\omega))$. Clearly $\lim_\alpha \gamma_\alpha(t) = \gamma(t)$ for each $t \in G$. We obtain a corresponding net $(\widetilde{\gamma}_\alpha)$ in $A_\omega \subset B_\omega$. By passing to a subnet, if necessary, we may suppose that $\widetilde{\gamma}_\alpha \to \Phi$ for some $\Phi \in B_\omega$. Let $t \in G$. Then

$$(\Phi \cdot \lambda)(t) = \langle \Phi, \lambda \cdot \delta_t \rangle = \lim_\alpha \langle \widetilde{\gamma}_\alpha, \lambda \cdot \delta_t \rangle = \lim_\alpha \gamma_\alpha(t) = \gamma(t) , \tag{11.3}$$

and so $\Phi \cdot \lambda = \gamma$. Since X is left-introverted, we conclude that $\gamma \in X$. Thus $\overline{\langle K_\lambda \rangle} \subset X$.

Conversely, suppose that $\overline{\langle K_\lambda \rangle} \subset X$ for each $\lambda \in X$. Then X is translation-invariant. Take $\Phi \in S_\omega$ and $\lambda \in X$. Since $S_\omega = \overline{\langle \Delta_\omega \rangle}$, there is a net (f_α) contained in $\langle \Delta_\omega \rangle$ such that $f_\alpha \to \Phi$. We may suppose that $f_\alpha = \widetilde{\gamma}_\alpha$ for each α, where (γ_α) is a net in $\langle K_\lambda \rangle$. By passing to a subnet, we may suppose that $\gamma_\alpha \to \gamma$ for some $\gamma \in A'_\omega$, and $\gamma \in \overline{\langle K_\lambda \rangle} \subset X$. The calculation in (11.3) shows that $\Phi \cdot \lambda = \gamma$, and so $\Phi \cdot \lambda \in X$. Thus X is left-introverted. □

In particular, each weak-$*$ closed, left-translation invariant subspace of A'_ω is left-introverted.

In Definition 5.5, we defined the topological centre $\mathfrak{Z}_t(X')$ of the dual X' of a left-introverted subset X of A'_ω. We now characterize $\mathfrak{Z}_t(X')$. Recall that $\Phi \in \mathfrak{Z}_t(X')$ if and only if the map $\Psi \mapsto \langle \Phi, \Psi \cdot \lambda \rangle$ is continuous on X' for each $\lambda \in X$.

PROPOSITION 11.2. *Let ω be a weight on a group G, let X be a left-introverted subspace of A'_ω, and let $\Phi \in X'$. Then $\Phi \in \mathfrak{Z}_t(X')$ if and only if*

$$\langle \Phi, \Psi_\alpha \cdot \lambda \rangle \to \langle \Phi, \Psi \cdot \lambda \rangle \quad (\lambda \in X) \tag{11.4}$$

whenever (Ψ_α) is a net in S_ω such that $\Psi_\alpha \to \Psi$ in $(X', \sigma(X', X))$.

PROOF. Let $\Phi \in \mathfrak{Z}_t(X')$. Then it is immediate that the specified condition is satisfied.

For the converse, take $\lambda \in X$. Then $\langle a, \lambda \cdot \Phi \rangle = \langle \Phi, a \cdot \lambda \rangle$ $(a \in A)$, and so, in particular,

$$\langle \Psi, \lambda \cdot \Phi \rangle = \langle \Phi, \Psi \cdot \lambda \rangle \tag{11.5}$$

for each $\Psi \in \Delta_\omega$, and hence for each $\Psi \in \langle \Delta_\omega \rangle \subset S_\omega$.

Now suppose that the specified condition holds, and take $\Psi \in S_\omega$. Then $\Psi = \lim_\alpha \Psi_\alpha$ for some net (Ψ_α) contained in $\langle \Delta_\omega \rangle$. Certainly we have immediately that $\langle \Psi, \lambda \cdot \Phi \rangle = \lim_\alpha \langle \Psi_\alpha, \lambda \cdot \Phi \rangle$, and, by the hypothesis, $\langle \Phi, \Psi \cdot \lambda \rangle = \lim_\alpha \langle \Phi, \Psi_\alpha \cdot \lambda \rangle$. Thus equation (11.5) holds for all $\Psi \in S_\omega$. Since $\operatorname{lin} S_\omega = B_\omega$, (11.5) holds for all $\Psi \in B_\omega$.

Let (Ψ_α) be any net in X' such that $\Psi_\alpha \to \Psi$ in X', and take $\lambda \in X$. Then

$$\langle \Phi, \Psi_\alpha \cdot \lambda \rangle = \langle \Psi_\alpha, \lambda \cdot \Phi \rangle \to \langle \Psi, \lambda \cdot \Phi \rangle = \langle \Phi, \Psi \cdot \lambda \rangle,$$

and so $\Phi \in \mathfrak{Z}_t(X')$, as required. □

PROPOSITION 11.3. *Let ω be a weight on a group G, and suppose that Ω clusters on $G \times G$. Then each $\|\cdot\|$-closed, translation-invariant subspace of A'_ω which is contained in $WAP(G, 1/\omega)$ is introverted.*

PROOF. Let X be $\|\cdot\|$-closed, translation-invariant subspace of A'_ω such that $X \subset WAP(G, 1/\omega)$, and take $\lambda \in X$. As in (11.2), set $K_\lambda = \left\{ \widetilde{\delta_s} \cdot \lambda : s \in G \right\}$.

The map

$$T : \lambda \mapsto \lambda/\omega, \quad A'_\omega \to \ell^\infty(G),$$

is a linear isometry. Since $\lambda \in WAP(G, 1/\omega)$, we have $T\lambda \in WAP(G)$ by definition. Also, for each $s, t \in G$, we have

$$T(\widetilde{\delta_t} \cdot \lambda)(s) = (\delta_t \cdot \lambda)(s)/\omega(s)\omega(t) = \Omega(s,t)(T\lambda)(st).$$

By hypothesis, Ω clusters on $G \times G$. Also, by Theorem 6.3, the map

$$f : (s,t) \mapsto (T\lambda)(st), \quad G \times G \to \mathbb{C},$$

clusters on $G \times G$. Since both Ω and f are bounded on $G \times G$, it follows from a remark after Definition 3.2 that the function

$$(s,t) \mapsto T(\widetilde{\delta_t} \cdot \lambda)(s), \quad G \times G \to \mathbb{C},$$

also clusters on $G \times G$. By Theorem 3.3, the set

$$\left\{T(\widetilde{\delta_t} \cdot \lambda) : t \in G\right\}$$

is relatively weakly compact in $CB(G)$, and so K_λ is weakly compact in A'_ω. By the Krein–Šmulian theorem, $\langle K_\lambda \rangle$ is relatively weakly compact in A'_ω. Let H_λ be the weak closure of $\langle K_\lambda \rangle$. Then H_λ is weakly compact, and so $H_\lambda = \overline{\langle K_\lambda \rangle}$. By Mazur's theorem, H_λ is also the $\| \cdot \|$-closure of $\langle K_\lambda \rangle$. Since the set X is $\| \cdot \|$-closed in A'_ω, it follows that $\overline{\langle K_\lambda \rangle} \subset X$. By Lemma 11.1, X is left-introverted.

Similarly, X is right-introverted, and so X is an introverted subspace of A'_ω. □

THEOREM 11.4. *Let ω be a weight on a group G such that Ω clusters on $G \times G$, and let X be a left-introverted subspace of A'_ω with*

$$X \subset WAP(G, 1/\omega).$$

Then $\mathfrak{Z}_t(X') = X'$.

PROOF. Let $\Phi \in X'$, and let (Ψ_α) be a net in S_ω such that $\Psi_\alpha \to \Psi$ in $(X', \sigma(X', X))$. By Proposition 11.2, $\Phi \in \mathfrak{Z}_t(X')$ provided that (11.4) holds.

Take $\lambda \in X$. As in the proof of Proposition 11.3, the set $\overline{\langle K_\lambda \rangle}$ is weakly compact in A'_ω, and so the weak and pointwise topologies coincide on $\overline{\langle K_\lambda \rangle}$. Since $(\Psi_\alpha \cdot \lambda)(t) \to (\Psi \cdot \lambda)(t)$ for each $t \in G$, $\Psi_\alpha \cdot \lambda \to \Psi \cdot \lambda$ weakly in $\overline{\langle K_\lambda \rangle}$, so giving (11.4). □

We do not know whether or not the above theorem holds in the case where Ω does not cluster on $G \times G$; an example where Ω does not cluster was given in Example 9.12.

Let G be a group, and let X be a left-introverted subspace of ℓ^∞. Then $X \subset WAP(G)$ whenever $\mathfrak{Z}_t(X') = X'$, and so there is a converse to the above theorem in the special case where $\omega = 1$. To see this, take $\lambda \in X$. Since $\mathfrak{Z}_t(X') = X'$, the map

$$\Phi \mapsto \Phi \cdot \lambda, \quad (X', \sigma(X', X)) \to (X, \sigma(X, X')),$$

is continuous. Hence $\{\Phi \cdot \lambda : \Phi \in (X')_{[1]}\}$ is relatively weakly compact in X. But this latter set contains K_λ, so that the set $\{r_t\lambda : t \in G\}$ is relatively weakly compact, and hence $\lambda \in WAP(G)$. Thus we have $X \subset WAP(G)$.

However this converse is not necessarily true for a non-trivial weight ω on G. For set $X = \ell^1(\omega_\alpha)'$, where $\alpha > 0$. Then X is left-introverted and $\mathfrak{Z}_t(X') = X'$ because $\ell^1(\omega_\alpha)'$ is Arens regular by Example 9.1. On the other hand, $X \not\subset WAP(\mathbb{Z}, 1/\omega_\alpha)$ because $\ell^\infty \not\subset WAP(\mathbb{Z})$.

Now let ω be a weight function on a locally compact group, and again suppose throughout that $\omega(s) \geq 1$ $(s \in G)$. Our next aim is to prove that, under certain conditions on ω, we have $M(G, \omega) = \mathfrak{Z}_t(\mathcal{X}'_\omega)$, where we are again setting $\mathcal{X}_\omega = LUC(G, 1/\omega)$. In the case where $\omega = 1$, this result was proved for G a locally compact abelian group by M. Grosser and Losert in [GLos] and for arbitrary G in [L3], and the present proof develops the earlier arguments. Again, we are setting $\mathcal{M}_\omega = M(G, \omega)$ and $\mathcal{E}_\omega = C_0(G, 1/\omega)$. Recall from Theorem 7.19 that $\mathcal{X}_\omega$ is left-introverted as a subspace of both $\mathcal{A}'_\omega$ and A'_ω, from Proposition 7.20 that the product $\Box$ is independent of the ambient space, and from the remark above Theorem 7.25 that $\mathcal{M}_\omega$ is canonically a closed subalgebra of $\mathcal{X}'_\omega$.

We make some preliminary remarks. We regard each element

$$\widetilde{\delta}_s : \lambda \mapsto \lambda(s)/\omega(s), \quad \mathcal{X}_\omega \to \mathbb{C},$$

as a character on $\mathcal{X}_\omega$. The space $\mathcal{X}_\omega$ is a C^*-subalgebra of the commutative C^*-algebra $\mathcal{A}'_\omega$; we denote the character space of $\mathcal{X}_\omega$ by Δ_ω, as in Chapter 8, so that Δ_ω is a compact subspace of $(\mathcal{X}'_\omega)_{[1]}$. The map

$$s \mapsto \widetilde{\delta}_s, \quad G \to \Delta_\omega,$$

is an embedding, and G is dense in Δ_ω, and so Δ_ω is a compactification of the locally compact space G. In the case where $\omega = 1$, the space Δ_ω is a semigroup; it is discussed in [LPy2], for example.

As in (7.21), we regard $\mathcal{M}_\omega$ as a closed subalgebra of $\mathfrak{Z}_t(\mathcal{X}'_\omega)$.

LEMMA 11.5. *Let ω be a weight function on a locally compact group G, and let $\lambda \in \mathcal{X}_\omega$ and $\Phi \in \mathfrak{Z}_t(\mathcal{X}'_\omega)$. Then $\lambda \cdot \Phi \in \mathcal{X}_\omega$.*

PROOF. Let $s_\alpha \to s$ in G. Then $\widetilde{\delta}_{s_\alpha} \to \widetilde{\delta}_s$ in $(\mathcal{X}'_\omega, \sigma(\mathcal{X}'_\omega, \mathcal{X}_\omega))$, and so

$$\begin{aligned}(\lambda \cdot \Phi)(s_\alpha)/\omega(s_\alpha) &= \langle \Phi, \widetilde{\delta}_{s_\alpha} \cdot \lambda\rangle = \langle \Phi \Box \widetilde{\delta}_{s_\alpha}, \lambda\rangle \\ &\to \langle \Phi \Box \widetilde{\delta}_s, \lambda\rangle = (\lambda \cdot \Phi)(s)/\omega(s)\end{aligned}$$

because L_Φ is continuous on $\mathcal{X}'_\omega$, and hence the function $(\lambda \cdot \Phi)/\omega$ is bounded and continuous on G.

Let $\Psi \in B_\omega$ and $s \in G$, and consider the equation

$$\langle \Psi, \lambda \cdot \Phi \cdot \delta_s \rangle = \langle \Phi \Box \delta_s \Box \Psi, \lambda \rangle . \tag{11.6}$$

This clearly holds in the case where $\Psi = \delta_t$ for $t \in G$ because both sides are equal to $\langle \Phi, \delta_{st} \cdot \lambda \rangle$, and so it holds whenever $\Psi \in \langle \Delta_\omega \rangle$. Now take $\Psi \in S_\omega$, say $\Psi = \lim_\alpha \Psi_\alpha$ for a net (Ψ_α) in $\langle \Delta_\omega \rangle$. Then

$$\lim_\alpha \langle \Phi \Box \delta_s \Box \Psi_\alpha, \lambda \rangle = \langle \Phi \Box \delta_s \Box \Psi, \lambda \rangle$$

again because L_Φ and δ_s are continuous on $\mathcal{X}_\omega'$. Thus (11.6) holds for each $\Psi \in S_\omega$, and hence for each $\Psi \in B_\omega$.

Assume towards a contradiction that $\lambda \cdot \Phi \notin \mathcal{X}_\omega$. By Proposition 7.15, the map $s \mapsto (\lambda \cdot \Phi) \cdot \delta_s$ is not continuous, and so there exists $s \in G$, $\varepsilon > 0$, and a net (s_α) in G with $s_\alpha \to s$ in G such that

$$\|(\lambda \cdot \Phi) \cdot \delta_{s_\alpha} - (\lambda \cdot \Phi) \cdot \delta_s\| > \varepsilon \quad \text{for each } \alpha .$$

Take $(\Psi_\alpha) \subset (\mathcal{X}_\omega')_{[1]}$ with

$$|\langle \Psi_\alpha, (\lambda \cdot \Phi) \cdot (\delta_{s_\alpha} - \delta_s) \rangle| > \varepsilon \quad \text{for each } \alpha .$$

By (11.6), we have

$$|\langle \Phi \Box (\delta_{s_\alpha} - \delta_s) \Box \Psi_\alpha, \lambda \rangle| > \varepsilon \quad \text{for each } \alpha .$$

Let $\Psi \in (\mathcal{X}_\omega')_{[1]}$ be a weak-$\star$ accumulation point of the net (Ψ_α). Then

$$|\langle \Phi, (\delta_{s_\alpha} \Box \Psi_\alpha - \delta_s \Box \Psi) \cdot \lambda \rangle| + |\langle \Phi \Box \delta_s, (\Psi - \Psi_\alpha) \cdot \lambda \rangle| > \varepsilon \tag{11.7}$$

for each α. However $\lim_\alpha \langle \Phi \Box \delta_s, (\Psi - \Psi_\alpha) \cdot \lambda \rangle = 0$. Also the map

$$\rho : (s, \Psi) \mapsto \delta_s \Box \Psi, \quad G \times (\mathcal{X}_\omega')_{[1]} \to \mathcal{X}_\omega' ,$$

is separately continuous, and so, by Ellis's theorem [El] (see also [Pa3, 12.1.5(a)] and [HiSt, Chapter 2.5]), ρ is continuous. Hence

$$\lim_\alpha \langle \Phi, (\delta_{s_\alpha} \Box \Psi_\alpha - \delta_s \Box \Psi) \cdot \lambda \rangle = 0 .$$

This is a contradiction of (11.7), and so $\lambda \cdot \Phi \in \mathcal{X}_\omega$, as required. □

PROPOSITION 11.6. *Let ω be a weight function on a locally compact group G, and let $\Phi \in \mathcal{X}_\omega'$. Then*

$$\Phi \in \mathfrak{Z}_t(\mathcal{X}_\omega') \quad \text{if and only if} \quad \lambda \cdot \Phi \in \mathcal{X}_\omega \;\; (\lambda \in \mathcal{X}_\omega)$$

and

$$\langle \Phi \Box \Psi, \lambda \rangle = \langle \Psi, \lambda \cdot \Phi \rangle \quad (\lambda \in \mathcal{X}_\omega, \Psi \in \mathcal{X}_\omega') . \tag{11.8}$$

PROOF. Let $\Phi \in \mathfrak{Z}_t(\mathcal{X}'_\omega)$. Then $\lambda \cdot \Phi \in \mathcal{X}_\omega$ $(\lambda \in \mathcal{X}_\omega)$ by Lemma 11.5.

Let $\lambda \in \mathcal{X}_\omega$. Equation (11.8) certainly holds in the special case where $\Psi = \delta_t$ for some $t \in G$, and so it holds whenever $\Psi \in \langle \Delta_\omega \rangle$. Now take $\Psi \in S_\omega$, say $\Psi = \lim_\alpha \Psi_\alpha$ for a net (Ψ_α) in $\langle \Delta_\omega \rangle$. Then $\lim_\alpha \langle \Phi \,\Box\, \Psi_\alpha, \lambda \rangle = \langle \Phi \,\Box\, \Psi, \lambda \rangle$ because L_Φ is continuous on $\mathcal{X}'_\omega$, and certainly $\lim_\alpha \langle \Psi_\alpha, \lambda \cdot \Phi \rangle = \langle \Psi, \lambda \cdot \Phi \rangle$, and so (11.8) holds for each $\Psi \in S_\omega$, and hence for each $\Psi \in B_\omega$.

The converse is trivial. □

The following is a key lemma; it is the point where we use the fact that Φ is in the topological centre of $\mathcal{X}'_\omega$.

LEMMA 11.7. *Let $\Phi \in \mathfrak{Z}_t(\mathcal{X}'_\omega)$ with $\|\Phi\| \leq 1$. Then the operator Φ_r belongs to the strong-operator closure of* $\mathrm{ac}\,\{\ell_s/\omega(s) : s \in G\}$.

PROOF. Let $\Phi \in (\mathcal{X}'_\omega)_{[1]}$. There is a net (f_α) in the set

$$\mathrm{ac}\left\{\widetilde{\delta}_s : s \in G\right\}$$

with $f_\alpha \to \Phi$ in $(\mathcal{X}'_\omega, \sigma(\mathcal{X}'_\omega, \mathcal{X}_\omega))$. Suppose, further, that $\Phi \in \mathfrak{Z}_t(\mathcal{X}'_\omega)$. Then

$$\begin{aligned}\lim_\alpha \langle \Psi, \lambda \cdot f_\alpha \rangle &= \lim_\alpha \langle f_\alpha \,\Box\, \Psi,\, \lambda \rangle = \langle \Phi \,\Box\, \Psi,\, \lambda \rangle \\ &= \langle \Phi \Diamond \Psi,\, \lambda \rangle = \langle \Psi, \lambda \cdot \Phi \rangle \quad (\lambda \in \mathcal{X}_\omega,\ \Psi \in \mathcal{X}'_\omega)\end{aligned}$$

because R_Ψ is continuous on $\mathcal{X}_\omega$. This says that the operator Φ_r of $\mathcal{B}(\mathcal{X}_\omega)$ is in the weak-operator closure of the set $\mathrm{ac}\{\ell_s/\omega(s) : s \in G\}$. By a theorem of Bade [DfS, Corollary VI.1.5], the weak-operator and strong-operator closures of this set are equal. □

Recall that $\kappa(G)$ was defined in Definition 7.39.

DEFINITION 11.8. *Let Ω be a locally compact space, and let S be a subset of Ω. Then S is* dispersed *if $S \not\subset \bigcup\{L_j : j \in J\}$ for any family $\{L_j : j \in J\}$ of compact sets with $|J| < \kappa(G)$.*

A subset S of a discrete space Ω is dispersed if and only if $|S| = |\Omega|$; a subset S of $\mathbb{R}$ dispersed if and only if S is unbounded.

The proof of the following theorem is similar to that of [L3, Theorem 1], but some details are different, and so we give essentially the full argument.

THEOREM 11.9. *Let ω be a weight function on a locally compact group G. Suppose that ω is diagonally bounded on a dispersed subset of G. Then $\mathfrak{Z}_t(\mathcal{X}'_\omega) = M(G, \omega)$.*

PROOF. By (7.22), it suffices to show that $\mathfrak{Z}_t(\mathcal{X}'_\omega) \cap \mathcal{E}^\circ_\omega = \{0\}$.

First suppose that the group G is compact. Then $\mathcal{X}_\omega = C(G)$ and $\mathcal{M}_\omega = M(G) = \mathcal{X}'_\omega$, and so the result is immediate. Henceforth we suppose that G is not compact.

Take $\Phi \in \mathfrak{Z}_t(\mathcal{X}'_\omega) \cap \mathcal{E}^\circ_\omega$ with $\|\Phi\| \leq 1$, and assume towards a contradiction that $\Phi \neq 0$. Thus we may suppose that $\|\Phi\| = 1$.

Let $\{K_i : i \in I\}$ be a family of compact subsets of G such that $|I| = \kappa(G)$ and $\bigcup\{K_i : i \in I\} = G$; we may also suppose that this family is closed under finite unions.

Let V be a compact neighbourhood of e_G, and set $\varphi = \chi_V/m(V)$ (where m denotes the left Haar measure on G). Define a pseudometric d_φ on G by the formula

$$d_\varphi(s,t) = \|\varphi \star \delta_s - \varphi \star \delta_t\| \quad (s,t \in G)\,.$$

Now choose $\varepsilon > 0$ such that $\varepsilon < 1/6$. Then there exists $\lambda \in (\mathcal{X}_\omega)_{[1]}$ with $|\langle \Phi, \lambda\rangle| > 1 - \varepsilon$. Set $f = \lambda/\omega$, so that $f \in LUC(G)$. For each $i \in I$, we define

$$f_i(s) = (1 - \min\{1, d_\varphi(s, K_i)\})f(s) \quad (s \in G)\,,$$

so that $f_i \in LUC(G)$ and $f_i(s) = f(s)$ $(s \in K_i)$, and we also define

$$\begin{aligned} B_i &= \{s \in G : d_\varphi(s, K_i) \leq 1\}\,,\\ A_i &= \{s \in G : d_\varphi(s, B_i) \leq 1\}\,, \end{aligned}$$

so that $K_i \subset B_i \subset A_i \subset V^{-1} \cdot (V \cdot B_i)$, the closed set A_i is compact, and $\operatorname{supp} f_i \subset B_i$.

Let S be a dispersed subset of G such that ω is diagonally bounded on S. For each $i \in I$, choose elements u_i and v_i in S such that each of the two families $\{A_i \cdot u_i : i \in I\}$ and $\{A_i \cdot v_i : i \in I\}$ is pairwise disjoint and such that

$$(11.9) \qquad \left(A_i \cdot u_i u_j^{-1}\right) \cap \left(A_k \cdot v_k v_\ell^{-1}\right) = \emptyset$$

whenever $i, j, k, \ell \in I$ with $i \neq j$ and $k \neq \ell$. (For details of the choice of these elements, see [L2, Lemma 1]; it is at this point that we use the fact that the set S is dispersed.)

We now define two functions λ' and λ'' on G.

For each $s \in A_i$, we set

$$(11.10) \qquad \left\{ \begin{aligned} \lambda'(su_i) &= \omega(s)f_i(s)/\omega(u_i^{-1})\,,\\ \lambda''(sv_i) &= \omega(s)f_i(s)/\omega(v_i^{-1})\,, \end{aligned} \right.$$

and we take $\lambda'(t) = \lambda''(t) = 0$ at points $t \in G$ at which these functions are not otherwise defined. It follows that λ' and λ'' are both well-defined.

We *claim* that $\lambda' \in \mathcal{X}_\omega$.

To see this claim, first take $t \in A_i \cdot u_i$. Then

$$|\lambda'(t)| \leq \omega(tu_i^{-1}) \left|f_i(tu_i^{-1}\right| / \omega(u_i^{-1}) \leq \omega(t) \left|f_i\right|_G \leq \omega(t) \left|f\right|_G ,$$

and this implies that $\lambda' \in \ell^\infty(G, 1/\omega)$.

Next, let (s_α) be a net such that $s_\alpha \to s$ in G, and fix δ with $0 < \delta < 1$. Then there exists α_0 such that

$$|\omega(s_\alpha) - \omega(s)| < \delta/(|f|_\Omega + 1), \quad d_\varphi(s_\alpha, s) < \delta,$$
$$|f(s_\alpha t) - f(st)| < \delta \quad (t \in G)$$

for each $\alpha \succ \alpha_0$.

We must consider three cases.

1) Suppose that $st \in B_i \cdot u_i$ for some (necessarily unique) $i \in I$. Then $stu_i^{-1} \in B_i$, and so $s_\alpha tu_i^{-1} \in A_i$ $(\alpha \succ \alpha_0)$. Hence, for each $\alpha \succ \alpha_0$, we have

$$|\lambda'(s_\alpha t) - \lambda'(st)| / \omega(t) \;=$$
$$\left| \frac{\omega(s_\alpha tu_i^{-1})}{\omega(s_\alpha)\omega(u_i^{-1})} \omega(s_\alpha) f_i(s_\alpha tu_i^{-1}) - \frac{\omega(stu_i^{-1})}{\omega(s)\omega(u_i^{-1})} \omega(s) f_i(stu_i^{-1}) \right| \Big/ \omega(t) ,$$

where we note that

$$|f_i(s) - f_i(t)| \leq |f(s) - f(t)| + d_\varphi(s,t) \quad (s,t \in G) .$$

2) Suppose that $st \in (A_i \cdot u_i) \setminus (B_i \cdot u_i)$ for some $i \in I$. For each $j \in I$ with $j \neq i$, necessarily $s_\alpha t \notin B_j \cdot u_j$ $(\alpha \succ \alpha_0)$. If $\alpha \succ \alpha_0$ and $s_\alpha t \notin B_i \cdot u_i$, then

$$|(\lambda'/\omega)(s_\alpha t) - (\lambda'/\omega)(st)| = 0 .$$

If $\alpha \succ \alpha_0$ and $s_\alpha t \in B_i \cdot u_i$, then $|(\lambda'/\omega)(s_\alpha t) - (\lambda'/\omega)(st)| < 3\delta$, as in 1).

3) Suppose that $st \notin A_i \cdot u_i$ for any $i \in I$. Then, for each $\alpha \succ \alpha_0$, necessarily $s_\alpha t \notin B_i \cdot u_i$ for any $i \in I$, and so

$$|(\lambda'/\omega)(s_\alpha t) - (\lambda'/\omega)(st)| = 0 .$$

In each of the three cases, we have shown that

$$|(\lambda'/\omega)(s_\alpha t) - (\lambda'/\omega)(st)| < 3\delta$$

for each $t \in G$ and each $\alpha \succ \alpha_0$. Hence $\|\lambda' \cdot \delta_{s_\alpha} - \lambda' \cdot \delta_s\| \to 0$. It follows from Proposition 7.15 that $\lambda' \in \mathcal{X}_\omega$, as claimed.

Similarly $\lambda'' \in \mathcal{X}_\omega$.

Since ω is diagonally bounded on S, there exists $m > 0$ such that

$$\omega(s)\omega(s^{-1}) \leq m \quad (s \in S) .$$

By Lemma 11.7, which applies because $\Phi \in \mathfrak{Z}_t(\mathcal{X}'_\omega)$, Φ_r belongs to the strong-operator closure of the set $\mathrm{ac}\{\ell_s/\omega(s) : s \in G\}$. Thus there exist $n \in \mathbb{N}$, numbers $c_1, \dots, c_n \in \mathbb{C}$ with $\sum_{j=1}^n |c_j| \leq 1$, and $s_1, \dots, s_n \in G$, such that

$$\left\{ \begin{array}{l} \|T(\lambda') - \lambda' \cdot \Phi\| < \varepsilon/m\,, \\ \|T(\lambda'') - \lambda'' \cdot \Phi\| < \varepsilon/m\,, \end{array} \right. \tag{11.11}$$

where $T = \sum_{j=1}^n c_j \ell_{s_j}/\omega(s_j) \in \mathcal{B}(\mathcal{X}_\omega)$. We may also suppose that

$$|\langle \mu, \lambda \rangle - \langle \Phi, \lambda \rangle| < \varepsilon,$$

where $\mu = \sum_{j=1}^n c_j \widetilde{\delta}_{s_j}$.

There exists $i \in I$ such that $s_1, \dots, s_n \in K_i$. Define

$$\gamma'(s) = \omega(u_i^{-1})\lambda'(su_i), \quad \gamma''(s) = \omega(v_i^{-1})\lambda''(sv_i) \quad (s \in G).$$

Thus $|\gamma'(s)| \vee |\gamma''(s)| \leq \omega(s)\,|f(s)|$ $(s \in G)$ and

$$\gamma'(s) = \gamma''(s) = \omega(s)f_i(s) \quad (s \in K_i).$$

We make the following calculation:

$$\begin{aligned} |(T\lambda')(u_i) - (\lambda' \cdot \Phi)(u_i)| &= \left| \sum_{j=1}^n c_j \lambda'(s_j u_i)/\omega(s_j) - \langle \Phi, \delta_{u_i} \cdot \lambda' \rangle \right| \\ &= \left| \sum_{j=1}^n c_j f(s_j)/\omega(u_i^{-1}) - \langle \Phi, \gamma'/\omega(u_i^{-1}) \rangle \right| \\ &= |\langle \mu, \lambda \rangle - \langle \Phi, \gamma' \rangle| \,/ \omega(u_i^{-1})\,. \end{aligned}$$

But $|(T\lambda')(u_i) - (\lambda' \cdot \Phi)(u_i)| < \varepsilon\omega(u_i)/m$ by (11.11), and so

$$|\langle \mu, \lambda \rangle - \langle \Phi, \gamma'' \rangle| \leq \varepsilon\omega(u_i^{-1})\omega(u_i)/m \leq \varepsilon\,.$$

However $|\langle \Phi, \lambda \rangle| > 1 - \varepsilon$ and $|\langle \mu, \lambda \rangle - \langle \Phi, \lambda \rangle| < \varepsilon$, and hence we have $|\langle \Phi, \gamma' \rangle| > 1 - 3\varepsilon$. Similarly, we have $|\langle \Phi, \gamma'' \rangle| > 1 - 3\varepsilon$. Since

$$\operatorname{supp} \gamma' \subset \bigcup_{j \in I} \{A_j \cdot u_j u_i^{-1}\} \quad \text{and} \quad \operatorname{supp} \gamma'' \subset \bigcup_{k \in I} \{A_k \cdot u_k u_i^{-1}\}\,,$$

it follows from (11.9) that

$$(\operatorname{supp} \gamma') \cap (\operatorname{supp} \gamma'') = \operatorname{supp}(\gamma' \cdot_\omega \gamma_i'') \subset K_i\,.$$

Take $h \in C_{00}(G)$ to be such that $h(t) = 1$ $(t \in V \cup K_i)$. Then

$$\gamma' \cdot_\omega \gamma'' \cdot_\omega (wh) = \gamma' \cdot_\omega \gamma''\,.$$

Since $\Phi \in \mathcal{E}_\omega^\circ$, it follows that

$$\left\{ \begin{array}{l} |\langle \gamma' \cdot_\omega (\omega - \omega h), \Phi\rangle| > 1 - 3\varepsilon\,, \\ |\langle \gamma'' \cdot_\omega (\omega - \omega h), \Phi\rangle| > 1 - 3\varepsilon\,. \end{array} \right. \tag{11.12}$$

Let $|\Phi|$ be the absolute value of the functional Φ in $\mathcal{X}'_\omega$, so that we have $\|\,|\Phi|\,\| \le 1$. Then it follows from (11.12) that

$$\left\{ \begin{array}{l} |\gamma' \cdot_\omega (\omega - \omega h)|\,, |\Phi|\rangle > 1 - 3\varepsilon\,, \\ \langle|\gamma'' \cdot_\omega (\omega - \omega h)|\,, |\Phi|\rangle > 1 - 3\varepsilon\,. \end{array} \right. \tag{11.13}$$

(See [Ta, p. 140].) Since $\varepsilon < 1/6$, it follows that

$$\langle|\gamma' + \gamma''|\,|\omega - \omega h|\,, |\Phi|\rangle > 2 - 6\varepsilon > 1\,. \tag{11.14}$$

Since $(\operatorname{supp}\gamma') \cap (\operatorname{supp}\gamma'') \subset K_i$, since $\omega(s) - (\omega h)(s) = 0 \;\; (s \in K_i)$, and since $|\gamma'(s)| \vee |\gamma''(s)| \le \omega(s)\,|f(s)| \;\; (s \in G)$, we have

$$|\gamma'(s) + \gamma''(s)|\,|\omega(s) - (\omega h)(s)| \le \omega(s) \quad (s \in G)\,.$$

Thus $\|\,|\gamma' + \gamma''|\;|\omega - \omega h|\,\| \le 1$, and so $\langle|\gamma' + \gamma''|\,|\omega - \omega h|\,, |\Phi|\rangle \le 1$, a contradiction of (11.14).

Thus we have shown that $\Phi = 0$ whenever $\Phi \in \mathfrak{Z}_t(\mathcal{X}'_\omega) \cap \mathcal{E}_\omega^\circ$. The result follows. □

The first corollary recovers Theorem 8.15.

COROLLARY 11.10. *Let G be a group, and let ω be a weight on G such that ω is diagonally bounded on S for some subset S of G with $|S| = |G|$. Then $\ell^1(G, \omega)$ is strongly Arens irregular.*

PROOF. It can be checked that, in the special case of the present corollary, we do not need the condition that $\omega(s) \ge 1 \;\; (s \in G)$. The theorem shows that $\ell^1(G, \omega)$ is left strongly Arens irregular.

The opposite algebra to $\ell^1(G, \omega)$ is $\ell^1(H, \check{\omega})$, where H is the opposite group to G, and this algebra is also strongly left Arens irregular because $\check{\omega}$ is diagonally bounded on S. Hence $\ell^1(G, \omega)$ is right strongly Arens irregular.

Thus $\ell^1(G, \omega)$ is strongly Arens irregular. □

Let G be a locally compact group, and set $A = L^1(G)$. Then we have $X = A' \cdot A = LUC(G)$ and $\mathfrak{Z}_t(X') = M(G)$ by Theorem 11.9, and also $\mathfrak{Z}_t^{(1)}(A'') = A$. This shows that we can have $\mathfrak{Z}_t^{(1)}(A'') \cap X' \subsetneq \mathfrak{Z}_t(X')$ in the circumstances of Proposition 5.9.

EXAMPLE 11.11. *There is an unbounded, symmetric weight ω on $\mathbb{R}$ such that $\omega(0) = 1$ and $\omega(t) > 1 \;\; (t \in \mathbb{R} \setminus \{0\})$, such that*

$$\liminf_{n \to \infty} \omega(n) < \infty\,,$$

but such that $\mathfrak{Z}_t(\mathcal{X}'_\omega) = M(G, \omega)$.

PROOF. Define η on $\mathbb{Z}$ as in Example 9.17, and extend η to $\mathbb{R}$ by requiring it to be linear on each interval $[n, n+1]$ for $n \in \mathbb{Z}$. Set $\omega = \exp \eta$ on $\mathbb{R}$. Then ω satisfies the conditions of Theorem 11.9 (with $S = \{2^k : k \in \mathbb{N}\}$), and so the result follows. $\square$

CHAPTER 12

The Second Dual of $L^1(G,\omega)$

We now move our considerations from the discrete Beurling algebras $\ell^1(G,\omega)$ to the algebras $L^1(G,\omega)$, where the locally compact group G is not discrete.

Throughout this chapter, G will be a locally compact group, and ω will be a weight function on G; we shall concentrate on the case where G is not discrete, and so $L^1(G,\omega)$ is not unital. We shall always suppose that $\omega(s) \geq 1 \ (s \in G)$. The basic case that we have in mind is that in which $G = (\mathbb{R}, +)$ and $\omega = \omega_\alpha$, where $\omega_\alpha(t) = (1 + |t|)^\alpha \ (t \in \mathbb{R})$ for $\alpha > 0$. We shall seek analogues of earlier results which were established for the algebras $\ell^1(\mathbb{Z},\omega)$; for comparison, we recall that the Beurling algebras $\ell^1(\mathbb{Z},\omega_\alpha)$ are Arens regular whenever $\alpha > 0$.

First, we shall prove that $L^1(G,\omega)$ is strongly Arens irregular whenever $\mathfrak{Z}_t(\mathcal{X}'_\omega) = M(G,\omega)$ (with a very minor condition on G). It follows that $L^1(G,\omega)$ is strongly Arens irregular whenever G contains a dispersed subset S such that ω is diagonally bounded on S.

Secondly, we shall consider weights that do not satisfy this latter condition, and in particular we shall consider the weights ω_α on $\mathbb{R}$ for $\alpha > 0$. There are many elements of $\mathcal{B}_{\omega_\alpha}$ which are not in the topological centre $\mathfrak{Z}_t(\mathcal{B}_{\omega_\alpha})$, and one might guess that $\mathcal{A}_{\omega_\alpha}$ is indeed strongly Arens irregular. However, we shall prove in Theorem 12.6 that this is not the case by exhibiting some specific elements of $\mathfrak{Z}_t(\mathcal{B}_{\omega_\alpha}) \setminus \mathcal{A}_{\omega_\alpha}$.

Finally, we shall identify the radical $\mathcal{R}_\omega^\Box$ of $\mathcal{B}_\omega = (L^1(G,\omega)'', \Box)$ in certain cases. The main result is Theorem 12.9, which is an analogue of Theorem 8.19. Subsequent results will elucidate the structure of $\mathcal{R}_\omega^\Box$.

We shall use notation introduced in (7.16). Thus $\mathcal{E}_\omega$ is a closed subspace of $\mathcal{A}'_\omega$, and $\mathcal{A}_\omega \subset \mathcal{E}'_\omega$, but it is no longer true in general that $\mathcal{A}_\omega = \mathcal{E}'_\omega$; indeed, $\mathcal{E}'_\omega = \mathcal{M}_\omega$ as Banach spaces, and $\mathcal{A}_\omega = \mathcal{M}_\omega$ only if G is discrete. Thus, as we stated in Chapter 4, $\mathcal{M}_\omega$ is a dual Banach algebra.

There are two general results that apply to the algebras $\mathcal{A}_\omega$. First, each has a bounded approximate identity, and so $\mathcal{B}_\omega = \mathcal{A}''_\omega$ has a (non-unique) mixed identity Φ_0. By Proposition 7.17(i), $\mathcal{A}'_\omega \cdot \mathcal{A}_\omega = \mathcal{X}_\omega$, and thus, by Proposition 5.9, we can write $(\mathcal{B}_\omega, \Box)$ as the semidirect product

$$\mathcal{B}_\omega = \Phi_0 \,\square\, \mathcal{B}_\omega \ltimes (\mathcal{A}'_\omega \cdot \mathcal{A}_\omega)^\circ = \mathcal{X}'_\omega \ltimes \mathcal{X}^\circ_\omega \,. \tag{12.1}$$

Let $\mathcal{Y}_\omega = (\Phi_0 \cdot \mathcal{B}_\omega) \cap \mathcal{E}^\circ_\omega$. Then $\mathcal{Y}_\omega$ is a closed subalgebra of $(\mathcal{B}_\omega, \square)$, with $\mathcal{Y}_\omega \oplus \mathcal{X}^\circ_\omega = \mathcal{E}^\circ_\omega$, and it follows from (12.1), (7.24), and (7.25) that

$$\mathcal{B}_\omega = \mathcal{A}_\omega \oplus \kappa_\omega(M_s(G,\omega)) \oplus \mathcal{Y}_\omega \oplus \mathcal{X}^\circ_\omega \tag{12.2}$$

as an ℓ^1-sum of Banach subspaces of $\mathcal{B}_\omega$. Here

$$\mathcal{A}_\omega \quad \text{and} \quad \mathcal{A}_\omega \oplus \kappa_\omega(M_s(G,\omega))$$

are closed subalgebras and $\mathcal{X}^\circ_\omega$ and $\mathcal{Y}_\omega \oplus \mathcal{X}^\circ_\omega$ are closed ideals in $\mathcal{B}_\omega$.

We next determine what can be said about $\mathfrak{Z}^{(1)}_t(\mathcal{B}_\omega)$ and $\mathfrak{Z}^{(2)}_t(\mathcal{B}_\omega)$ for an arbitrary weight function ω.

Let ω be a weight function on a non-discrete, locally compact group G. We consider the result that $\mathcal{A}_\omega$ is not Arens regular. This was first proved (in a stronger form) by Craw and Young [CrY, Theorem 2]: there is no weight function ω on G such that $\mathcal{A}_\omega$ is Arens regular. An alternative approach is to use Theorem 2.22. Indeed, $\mathcal{A}_\omega$ always has a bounded approximate identity and, by Theorem 7.1, $\mathcal{A}_\omega$ is weakly sequentially complete as a Banach space; also $\mathcal{A}_\omega$ does not have an identity, and so, by Theorem 2.22, $\mathcal{A}_\omega$ is not Arens regular. We seek to go further than these results by determining conditions under which $\mathcal{A}_\omega$ is strongly Arens irregular. Let Φ_0 be a mixed identity in $\mathcal{B}_\omega$. We can conclude from Theorem 2.21 that:

- $\mathfrak{Z}^{(2)}_t(\mathcal{B}_\omega) \subset \Phi_0 \,\square\, \mathcal{B}_\omega = \mathcal{X}'_\omega$;
- $\Phi_0 \notin \mathfrak{Z}^{(1)}_t(\mathcal{B}_\omega)$;
- $\mathcal{A}_\omega$ is not Arens regular.

For the proof of our main theorem, we require some preliminary remarks.

Let E be a Banach space, and let (x_k) be a sequence in E. The series $\sum_{k=1}^\infty x_k$ is *weakly unconditionally Cauchy* if $\sum_{k=1}^\infty |\langle x_k,\, \lambda\rangle| < \infty$ for each $\lambda \in E'$. In this case, the set $\{\sum_{k=1}^n x_k : n \in \mathbb{N}\}$ is weakly bounded in E, and so, by the uniform boundedness theorem, it is $\|\cdot\|$-bounded in E. Define

$$\langle \Lambda,\, \lambda\rangle = \sum_{k=1}^{\infty} \langle x,\, \lambda\rangle \quad (\lambda \in E')\,.$$

Then $\Lambda \in E''$ and $\Lambda = \lim_{n\to\infty} \sum_{k=1}^n x_k$ in $(E'', \sigma(E'', E'))$; we write this as $\Lambda = \sum_{k=1}^\infty x_k$. Now suppose that the series $\sum_{k=1}^\infty \lambda_k$ is weakly unconditionally Cauchy in E', and set

$$\Lambda = \sum_{k=1}^{\infty} \lambda_k \in E''' \quad \text{and} \quad \lambda = \Lambda \mid E\,.$$

Then $\lambda \in E'$ and $\lambda = \lim_{n\to\infty} \sum_{k=1}^{n} \lambda_k$ in $(E', \sigma(E', E))$; again we write $\lambda = \sum_{k=1}^{\infty} \lambda_k$.

A Banach space E has *Property* (X) if the following 'normality' condition holds: for each $\Lambda \in E''$ such that

$$\left\langle \Lambda, \sum_{k=1}^{\infty} \lambda_k \right\rangle = \sum_{k=1}^{\infty} \langle \Lambda,\, \lambda_k \rangle$$

whenever $\sum_{k=1}^{\infty} \lambda_k$ is weakly unconditionally Cauchy in E', it follows that $\Lambda \in E$. The space E has *Mazur's Property* if the following condition holds: for each $\Lambda \in E''$ such that $\lim_{n\to\infty}\langle \Lambda,\, \lambda_n \rangle = \langle \Lambda,\, \lambda \rangle$ whenever $\lim_{n\to\infty} \lambda_n = \lambda$ in $(E'_{[1]}, \sigma(E', E))$, it follows that $\Lambda \in E$.

The following results are taken from a clear and full account by Neufang in [N2]. Let ω be a weight function on a locally compact group G. Then $\mathcal{A}_\omega$ is the predual of the von Neumann algebra $\mathcal{A}'_\omega$, and so $\mathcal{A}_\omega$ has Mazur's property if and only if $\mathcal{A}_\omega$ has Property (X) [N2, Theorem 3.16]; further, $\mathcal{A}_\omega$ has these two equivalent properties if and only if $\kappa(G)$ is a non-measurable cardinal. (We recall that it cannot be proved in ZFC that measurable cardinals exist.) In particular, $\mathcal{A}_\omega$ has these properties whenever the group G is σ-compact. Indeed, it is remarked by Neufang in [N2] that the Banach space $\ell^1(\Gamma)$ has these properties if and only if $\kappa(\Gamma) = |\Gamma|$ is non-measurable, as proved by Edgar [Ed].

Recall from Proposition 7.17(ii) that, in the above notation, we have $\mathcal{A}_\omega \cdot \mathcal{A}'_\omega = RUC(G, 1/\omega)$.

LEMMA 12.1. *Let ω be a weight function on a locally compact group G for which $\kappa(G)$ is a non-measurable cardinal. Suppose that $\Phi \in \mathcal{B}_\omega$ has the following properties:*

(i) $f \cdot \Phi \in \mathcal{A}_\omega \;\; (f \in \mathcal{A}_\omega)$;

(ii) $\Phi \cdot \lambda \in \mathcal{A}_\omega \cdot \mathcal{A}'_\omega \;\; (\lambda \in \mathcal{A}'_\omega)$.

Then $\Phi \in \mathcal{A}_\omega$.

PROOF. Let $\sum_{k=1}^{\infty} \lambda_k$ be a weakly unconditionally Cauchy series in $\mathcal{A}'_\omega$, and set $\mu_n = \sum_{k=1}^{n} \lambda_k$ $(n \in \mathbb{N})$ and $\lambda = \lim_n \mu_n$, taking the limit in $(\mathcal{A}'_\omega, \sigma(\mathcal{A}'_\omega, \mathcal{A}_\omega))$.

By clause (ii), the sequence $(\Phi \cdot \mu_n : n \in \mathbb{N})$ is contained in the space $\mathcal{A}_\omega \cdot \mathcal{A}'_\omega$. Indeed, $(\Phi \cdot \mu_n : n \in \mathbb{N})$ is a bounded, weakly Cauchy sequence in $RUC(G, 1/\omega)$, and so

$$((\Phi \cdot \mu_n)(s) : n \in \mathbb{N}) = (\langle \delta_s, \Phi \cdot \mu_n \rangle : n \in \mathbb{N})$$

is a Cauchy sequence in $\mathbb{C}$ for each $s \in G$; we set

$$\mu(s) = \lim_n (\Phi \cdot \mu_n)(s) \quad (s \in G)\,.$$

Then μ/ω is a bounded, measurable function on G, and so $\mu \in \mathcal{A}'_\omega$.

Let $f \in \mathcal{A}_\omega$. Then

$$\langle f, \Phi \cdot \mu_n \rangle = \langle f \cdot \Phi, \mu_n \rangle \quad (n \in \mathbb{N})\,.$$

By clause (i), $f \cdot \Phi \in \mathcal{A}_\omega$, and so

$$\lim_n \langle f \cdot \Phi,\, \mu_n \rangle = \langle f \cdot \Phi,\, \lambda \rangle = \langle f,\, \Phi \cdot \lambda \rangle\,.$$

Thus $\Phi \cdot \mu_n \to \Phi \cdot \lambda$ in $(\mathcal{A}'_\omega, \sigma(\mathcal{A}'_\omega, \mathcal{A}_\omega))$. Further, for each $f \in \mathcal{A}_\omega$, we have

$$\langle f,\, \Phi \cdot \mu_n \rangle = \int_G f(s)(\Phi \cdot \mu_n)(s)\, \mathrm{d}m(s)\,.$$

Since $|(\Phi \cdot \mu_n)(s)| \leq C\omega(s)$ $(n \in \mathbb{N},\ s \in G)$ for some constant C, we can apply the dominated convergence theorem to see that

$$\langle f,\, \Phi \cdot \lambda \rangle = \int_G f(s)\mu(s)\, \mathrm{d}m(s) = \langle f,\, \mu \rangle\,.$$

Thus $\Phi \cdot \lambda = \mu$ locally almost everywhere on G.

By clause (ii), $\mu = \Phi \cdot \lambda \in \mathcal{A}_\omega \cdot \mathcal{A}'_\omega$, and so there exist $f \in \mathcal{A}_\omega$ and $\nu \in \mathcal{A}'_\omega$ such that $\mu = f \cdot \nu$. Let (e_α) be a bounded approximate identity for $\mathcal{A}_\omega$. Then

$$\lim_\alpha \langle e_\alpha \cdot \Phi,\, \lambda \rangle = \lim_\alpha \langle e_\alpha,\, f \cdot \nu \rangle = \lim_\alpha \langle e_\alpha \star f,\, \nu \rangle = \langle f,\, \nu \rangle\,.$$

This fact is exactly that which is required to deduce that μ is continuous locally almost everywhere at e_G, in the sense that, for each $\varepsilon > 0$ and each compact subset K of G, there is a neighbourhood V of e_G and a set N of measure 0 such that $|\mu(s) - \mu(e_G)| < \varepsilon$ for each $s \in V \cap K \setminus N$. This deduction is essentially that given in [IPyU, Lemma 2.3]; see also [LPy1, Lemma 2.14]. It follows easily that $\mu(e_G) = (\Phi \cdot \lambda)(e_G)$. We also have

$$(\Phi \cdot \lambda)(e_G) = \langle f,\, \nu \rangle = \langle \Phi_0 \cdot f,\, \nu \rangle = \langle \Phi_0,\, \Phi \cdot \lambda \rangle = \langle \Phi,\, \lambda \rangle\,,$$

where Φ_0 is a mixed unit. We conclude that $\mu(e_G) = \langle \Phi,\, \lambda \rangle$. In a similar way, we have $\mu_n(e_G) = (\Phi \cdot \lambda_n)(e_G)$ $(n \in \mathbb{N})$, and so we also have the equation $\mu_n(e_G) = \langle \Phi,\, \lambda_n \rangle$ $(n \in \mathbb{N})$. Thus

$$\langle \Phi,\, \lambda \rangle = \lim_n \langle \Phi,\, \mu_n \rangle = \sum_{k=1}^{\infty} \langle \Phi,\, \lambda_k \rangle\,.$$

Since $\kappa(G)$ is non-measurable, the Banach space $\mathcal{A}_\omega$ has Property (X), and so it follows that $\Phi \in \mathcal{A}_\omega$, as required. □

THEOREM 12.2. *Let ω be a weight function on a locally compact group G. Suppose that $\kappa(G)$ is a non-measurable cardinal and that $\mathfrak{Z}_t(\mathcal{X}'_\omega) = \mathcal{M}_\omega$. Then $L^1(G,\omega)$ is left strongly Arens irregular.*

PROOF. Let $\Phi \in \mathfrak{Z}_t^{(1)}(\mathcal{B}_\omega)$, and denote by $\widetilde{\Phi}$ the restriction of Φ to $\mathcal{X}_\omega$, so that $\widetilde{\Phi} \in \mathfrak{Z}_t(\mathcal{X}'_\omega)$.

Let $f \in \mathcal{A}_\omega$. Then $f \cdot \widetilde{\Phi} = f \cdot \Phi$ because $\mathcal{X}_\omega = \mathcal{A}'_\omega \cdot \mathcal{A}_\omega$. Since $\mathfrak{Z}_t(\mathcal{X}'_\omega) = \mathcal{M}_\omega$ by our hypothesis, we have $f \cdot \widetilde{\Phi} \in \mathcal{A}_\omega$. Hence $f \cdot \Phi \in \mathcal{A}_\omega$.

Now fix $g \in \mathcal{A}_\omega$, and set $\Psi = \Phi \cdot g$. Then $f \cdot \Psi \in \mathcal{A}_\omega \ (f \in \mathcal{A}_\omega)$ and $\Psi \cdot \lambda \in \mathcal{A}_\omega \cdot \mathcal{A}'_\omega \ (\lambda \in \mathcal{A}'_\omega)$. We have shown that Ψ satisfies the two conditions on Φ in Lemma 12.1, and so $\Psi \in \mathcal{A}_\omega$.

The analogue of Lemma 12.1 holds 'on the other side'. We know that $\Phi \cdot g \in \mathcal{A}_\omega \ (g \in \mathcal{A}_\omega)$, and $\lambda \cdot \Phi \in \mathcal{A}'_\omega \cdot \mathcal{A}_\omega \ (\lambda \in \mathcal{A}'_\omega)$ by Proposition 2.20. Thus, by this analogue, $\Phi \in \mathcal{A}_\omega$.

We have shown that $\mathfrak{Z}_t^{(1)}(\mathcal{B}_\omega) \subset \mathcal{A}_\omega$, and so $\mathcal{A}_\omega$ is left strongly Arens regular. □

THEOREM 12.3. *Let ω be a weight function on a locally compact group G. Suppose that ω is diagonally bounded on a dispersed subset of G, and suppose that $\kappa(G)$ is non-measurable. Then $L^1(G,\omega)$ is strongly Arens irregular.*

PROOF. It follows from Theorems 11.9 and 12.2 that $\mathcal{A}_\omega$ is left strongly Arens irregular. By applying this result to the opposite algebra, we see that $\mathcal{A}_\omega$ is right strongly Arens irregular. Thus $\mathcal{A}_\omega$ is strongly Arens irregular. □

We remark that Neufang has now proved the above theorem for each locally compact group in [N5] by a different method. In fact, a direct modification of the proof of Theorem 11.9 also proves the above theorem for each locally compact group G.

We now investigate the Beurling algebras $L^1(G,\omega)$ in the case where G is not discrete and ω does not satisfy the 'diagonal boundedness' condition of Theorem 12.3. Our aim is to show that $L^1(G,\omega)$ need not be strongly Arens irregular, and so some condition on ω is required in Theorem 12.3. In fact, we shall take G to be $(\mathbb{R},+)$, and work with weights ω on $\mathbb{R}$ such that Ω 0-clusters locally uniformly on $\mathbb{R}\times\mathbb{R}$ (see

Definition 3.7(ii)); for example, as we remarked, the weights ω_α on $\mathbb{R}$ have this property whenever $\alpha > 0$.

We first construct a special sequence of functions on $\mathbb{R}$.

Let (a_n) and (b_n) be sequences in $\mathbb{R}^+$ such that

$$a_1 < b_1 < a_2 < b_2 < \cdots ,$$

such that $r_n := b_n - a_n \to \infty$ and $a_{n+1} - b_n \to \infty$ as $n \to \infty$. Set $L_n = [a_n, b_n]$ $(n \in \mathbb{N})$. Then (g_n) is defined by

$$r_n g_n = \chi_{L_n} \quad (n \in \mathbb{N}) .$$

We see that $g_n \in L^1(\mathbb{R})$ and

$$\int_{\mathbb{R}} g_n(s)\,\mathrm{d}s = \|g_n\|_1 = 1 \quad (n \in \mathbb{N}) .$$

Let ω be a weight function on $\mathbb{R}$, and set $\widetilde{g_n} = g_n/\omega$ $(n \in \mathbb{N})$. Then $(\widetilde{g_n})$ is a sequence in $\mathcal{A}_\omega = L^1(\mathbb{R}, \omega)$; by passing to a subnet, we obtain a net $(\widetilde{g}_\beta)$ which converges in $\mathcal{B}_\omega$, say

$$\lim_\beta \widetilde{g}_\beta = \Psi_0 \in \mathcal{B}_\omega .$$

Note that $\Psi_0 \in \mathcal{E}^\circ_\omega$, but that $\Psi_0 \notin \mathcal{X}^\circ_\omega$ because $\langle \Psi_0, \omega\rangle = 1$. For each β, there exists $n_\beta \in \mathbb{N}$ such that $\widetilde{g}_\beta = \widetilde{g}_{n_\beta}$; we set $a_\beta = a_{n_\beta}$, $b_\beta = b_{n_\beta}$, $r_\beta = r_{n_\beta}$, and $L_\beta = L_{n_\beta}$, so that $\lim_\beta a_\beta = \lim_\beta r_\beta = \infty$ and $\operatorname{Lim}_\beta L_\beta = \infty$.

LEMMA 12.4. *Let ω be a weight function on $\mathbb{R}$, and let $\Phi \in \mathcal{B}_\omega$ be such that $\Phi = \lim_\alpha f_\alpha$ for a net (f_α) of continuous functions on $\mathbb{R}$ with $\|f_\alpha\|_\omega \le 1$ and* supp $f_\alpha \subset K$ *for each α, where K is a fixed compact subset of $\mathbb{R}$. Then there is a constant $z \in \overline{\mathbb{D}}$ such that*

$$\Phi \,\Box\, \Psi_0 = \Psi_0 \,\Box\, \Phi = z\Psi_0 .$$

PROOF. We may suppose that $K = [-1, 1]$. For each α, set

$$z_\alpha = \int_{-1}^{1} f_\alpha(s)\,\mathrm{d}s ,$$

so that $z_\alpha \in \overline{\mathbb{D}}$; clearly the net (z_α) converges, say $\lim_\alpha z_\alpha = z \in \overline{\mathbb{D}}$.

Take $\lambda \in L^\infty(\mathbb{R})$ with $\|\lambda\|_\infty \le 1$; we shall consider the number

$$x_{\alpha,\beta} := \langle f_\alpha \star \widetilde{g}_\beta, \lambda\omega\rangle ,$$

and compare it with the number

$$y_\beta := z\langle \widetilde{g}_\beta, \lambda\omega\rangle .$$

We see that, for each α and β, the functions $f_\alpha \star \widetilde{g}_\beta$ and $z_\alpha\widetilde{g}_\beta$ agree on the interval $[a_\beta+1, b_\beta-1]$ and that the support of $f_\alpha \star \widetilde{g}_\beta$ is contained in the interval $[a_\beta - 1, b_\beta + 1]$. Thus we have the estimate

$$|x_{\alpha,\beta} - y_\beta| \leq |z_\alpha - z| + 4C/r_\beta\,,$$

where $C = \sup\{\omega(u) : |u| \leq 1\}$. It follows that

$$\lim_\alpha \lim_\beta |x_{\alpha,\beta} - y_\beta| = \lim_\beta \lim_\alpha |x_{\alpha,\beta} - y_\beta| = 0\,.$$

Thus

$$\langle \Phi \,\Box\, \Psi_0\,, \lambda\omega\rangle = \lim_\alpha \lim_\beta x_{\alpha,\beta} = \lim_\beta y_\beta = z\langle \Psi_0, \lambda\omega\rangle$$

and

$$\langle \Psi_0 \,\Box\, \Phi, \,\lambda\omega\rangle = \lim_\beta \lim_\alpha x_{\alpha,\beta} = \lim_\beta y_\beta = z\langle \Psi_0, \lambda\omega\rangle\,.$$

These equalities hold for each $\lambda \in L^\infty(\mathbb{R})$, and so the result follows. □

LEMMA 12.5. *Let ω be a weight function on $\mathbb{R}$ such that Ω 0-clusters locally uniformly on $\mathbb{R}\times\mathbb{R}$, and let $\Phi \in \mathcal{B}_\omega$ be such that $\Phi = \lim_\alpha f_\alpha/\omega$ for a net (f_α) of continuous functions with compact support on $\mathbb{R}$ such that $\|f_\alpha\|_1 \leq 1$ for each α and such that $\mathrm{Lim}_{\,\alpha}\,\mathrm{supp}\ f_\alpha = \infty$. Then*

$$\Phi \,\Box\, \Psi_0 = \Psi_0 \,\Box\, \Phi = 0\,.$$

PROOF. Set $\widetilde{f_\alpha} = f_\alpha/\omega$ and $K_\alpha = \mathrm{supp}\ f_\alpha$ for each α.
Take $\lambda \in L^\infty(\mathbb{R})$ with $\|\lambda\|_\infty \leq 1$. Then we have the estimate

$$\begin{aligned}\left|\langle \widetilde{f_\alpha} \star \widetilde{g}_\beta,\, \lambda\omega\rangle\right| &\leq \int_{K_\alpha}\int_{L_\beta} |f_\alpha(s)|\,|g_\beta(s)|\,|\lambda(s+t)|\,\Omega(s,t)\,\mathrm{d}s\,\mathrm{d}t\\ &\leq \sup_{K_\alpha\times L_\beta}\ \Omega\,,\end{aligned}$$

and so

$$\lim_\alpha \lim_\beta \left|\langle \widetilde{f_\alpha} \star \widetilde{g}_\beta,\, \lambda\omega\rangle\right| = \lim_\beta \lim_\alpha \left|\langle \widetilde{f_\alpha} \star \widetilde{g}_\beta,\, \lambda\omega\rangle\right| = 0$$

because $\mathrm{Lim}_{\,\alpha}\, K_\alpha = \mathrm{Lim}_{\,\beta}\, L_\beta = \infty$ and Ω 0-clusters locally uniformly on $\mathbb{R}\times\mathbb{R}$. Thus

$$\langle \Phi \,\Box\, \Psi_0, \lambda\omega\rangle = \langle \Psi_0 \,\Box\, \Phi, \lambda\omega\rangle = 0\,.$$

These equalities hold for each $\lambda \in L^\infty(\mathbb{R})$, and so $\Phi \,\Box\, \Psi_0 = \Psi_0 \,\Box\, \Phi = 0$, as required. □

THEOREM 12.6. *Let ω be a weight function on $\mathbb{R}$ such that Ω 0-clusters locally uniformly on $\mathbb{R}\times\mathbb{R}$. Then the above element Ψ_0 of $\mathcal{B}_\omega$ belongs to the centre $\mathfrak{Z}(\mathcal{B}_\omega)$, and so $\mathcal{A}_\omega = L^1(\mathbb{R},\omega)$ is not strongly Arens irregular. Further, $\Psi_0 \in \mathcal{E}_\omega^\circ \setminus \mathcal{X}_\omega^\circ$.*

PROOF. Let $\Phi \in \mathcal{B}_\omega$ with $\|\Phi\| = 1$. Then there is a net (f_α) of continuous functions with compact support such that $\|f_\alpha\|_\omega = 1$ and $\lim_\alpha f_\alpha = \Phi$. For convenience, we suppose that $\operatorname{supp} f_\alpha \subset \mathbb{R}^+$ for each α, and work on $\mathbb{R}^+$.

Fix $\varepsilon > 0$. We shall show that $\|X\| < 4\varepsilon$, where

$$X = \Phi \,\Box\, \Psi_0 - \Psi_0 \,\Box\, \Phi\,.$$

This is sufficient to show that $\Psi_0 \in \mathfrak{Z}(\mathcal{B}_\omega)$, from which the result follows.

Let $\theta \in [0,1)$. For each α, consider the set

$$\left\{ s \in \mathbb{R}^+ : \int_0^s |f_\alpha(t)|\,\omega(t)\,\mathrm{d}t = \theta \right\}.$$

This set is clearly closed in $\mathbb{R}^+$ and non-empty; we specify $s(\alpha,\theta)$ to be its minimum. Note that, for each α, $s(\alpha,\theta)$ is an increasing function of θ. Define

$$c(\theta) = \sup_\alpha s(\alpha,\theta) \in \mathbb{R}^+ \cup \{\infty\}\,.$$

Then $c(\theta)$ is an increasing function of θ, and so

$$\{\theta \in [0,1) : c(\theta) < \infty\}$$

is a subinterval of $[0,1)$; it is non-empty because $c(0) = 0$. Let θ_0 be the supremum of this set, so that $\theta_0 \in [0,1]$.

We first suppose that $\theta_0 \in (0,1)$ and that $[\theta_0 - \varepsilon, \theta_0 + \varepsilon] \subset (0,1)$.

Set $c = c(\theta_0 - \varepsilon)$, and choose $\lambda \in C(\mathbb{R}^+)$ such that

$$\lambda(\mathbb{R}^+) \subset \mathbb{I}, \quad \lambda(s) = 1 \ (0 \le s \le c), \quad \text{and} \quad \operatorname{supp} \lambda \subset [0, c+1]\,.$$

Then set $\ell_\alpha = f_\alpha \lambda$ for each α, so that

$$\theta_0 - \varepsilon \le \|\ell_\alpha\|_\omega \le 1 \quad \text{for each } \alpha\,. \tag{12.3}$$

The set $\{s(\alpha, \theta_0 + \varepsilon)\}$ is unbounded, and so, by passing to a subnet of the original net (f_α), we may suppose that $\lim_\alpha t_\alpha = \infty$, where $t_\alpha = s(\alpha, \theta_0 + \varepsilon)$. For each α, choose $\rho_\alpha \in C(\mathbb{R}^+)$ such that

$$\rho_\alpha(\mathbb{R}^+) \subset \mathbb{I}, \quad \rho_\alpha(s) = 1 \ (s \ge t_\alpha), \quad \text{and} \quad \operatorname{supp} \rho_\alpha \subset (t_\alpha - 1, \infty)\,.$$

Then set $k_\alpha = f_\alpha \rho_\alpha$ for each α, so that

$$\theta_0 + \varepsilon \le \|k_\alpha\|_\omega \le 1 \quad \text{for each } \alpha\,. \tag{12.4}$$

By passing to a subnet of the original net (f_α), we may suppose that (ℓ_α) and (k_α) both converge in $\mathcal{B}_\omega$, say to Γ and Δ, respectively. By Lemma 12.4 (with $K = [0, c+1]$), we have $\Gamma \,\Box\, \Psi_0 = \Psi_0 \,\Box\, \Gamma$, and, by Lemma 12.5, we have $\Delta \,\Box\, \Psi_0 = \Psi_0 \,\Box\, \Delta = 0$, noting that $\operatorname{Lim}_\alpha \operatorname{supp} k_\alpha = \infty$ because $\operatorname{supp} \rho_\alpha \subset [t_\alpha - 1, \infty)$.

Take α such that $t_\alpha > c+2$. Then it follows from (12.3) and (12.4) that

$$\|f_\alpha - \ell_\alpha - k_\alpha\|_\omega \leq \int_c^{t_\alpha} |f_\alpha(s)|\,\omega(s)\,\mathrm{d}s \leq 2\varepsilon\,,$$

and so $\|\Phi - \Gamma - \Delta\| \leq 2\varepsilon$.

It now follows that $X = (\Phi - \Gamma - \Delta)\,\Box\,\Psi_0 - \Psi_0\,\Box\,(\Phi - \Gamma - \Delta)$, and so $\|X\| \leq 4\varepsilon$, as required.

It remains to consider the cases where $\theta_0 = 0$ or $\theta_0 = 1$. Suppose that $\theta_0 = 0$. Then the above argument applies without the need to introduce the function λ. Suppose that $\theta_0 = 1$. If there exists $c \in \mathbb{R}^+$ such that

$$\int_0^c |f_\alpha(t)|\,\omega(t)\,\mathrm{d}t = 1$$

for each sufficiently large α, then Lemma 12.4 applies directly; if there is no such constant c, the above argument applies without the need to introduce the functions ρ_α.

This concludes the proof in all cases. $\Box$

We have shown that the algebras $\mathcal{A}_\omega$ of the above theorem are neither Arens regular nor strongly Arens irregular, but we have no characterization of $\mathfrak{Z}(\mathcal{B}_\omega)$.

We now investigate the radicals of our second dual algebras.

PROPOSITION 12.7. *Let ω be a weight function on a locally compact group G, and suppose that $\mathcal{M}_\omega$ is semisimple. Then:*

(i) $\mathcal{R}_\omega^\Box \subset \mathcal{E}_\omega^\circ$*;*

(ii) $\mathcal{X}_\omega^\circ$ *is a closed, left-annihilator ideal of* $(\mathcal{B}_\omega,\,\Box\,)$*, and* $\mathcal{X}_\omega^\circ \subset \mathcal{R}_\omega^\Box$*;*

(iii) *for each mixed identity* Φ_0 *of* $\mathcal{B}_\omega$*,* $\Phi_0 + \mathcal{X}_\omega^\circ$ *is the identity of* $(\mathcal{X}_\omega',\,\Box\,)$.

PROOF. (i) This is immediate.

(ii) The result was discussed in Chapter 5.

(iii) There is a bounded approximate identity (e_α) in $\mathcal{A}_\omega$ such that $\lim_\alpha e_\alpha = \Phi_0$. Take $\Phi \in \mathcal{B}_\omega$ and $\lambda \in \mathcal{X}_\omega$. Then

$$\langle \Phi_0\,\Box\,\Phi,\ \lambda\rangle = \lim_\alpha \langle \Phi,\ \lambda \cdot e_\alpha\rangle = \langle \Phi,\ \lambda\rangle\,,$$

and so $\Phi_0\,\Box\,\Phi - \Phi \in \mathcal{X}_\omega^\circ$. Certainly $\Phi\,\Box\,\Phi_0 - \Phi = 0$. Since $(\mathcal{X}_\omega',\,\Box\,)$ is the quotient of $(\mathcal{B}_\omega,\,\Box\,)$ by the closed ideal $\mathcal{X}_\omega^\circ$, the result follows. $\Box$

We now seek to analyze the structure of $\mathcal{B}_\omega$ in the case where Ω 0-clusters strongly on $G \times G$.

LEMMA 12.8. *Suppose that Ω 0-clusters strongly on $G\times G$, and take $\Phi \in \mathcal{E}_\omega^\circ$ and $\lambda \in \mathcal{X}_\omega$. Then $\Phi \cdot \lambda \in \mathcal{E}_\omega$ and $\lambda \cdot \Phi \in \mathcal{E}_\omega$.*

PROOF. We may suppose that $\|\Phi\| = \|\lambda\| = 1$.

Fix $\varepsilon > 0$, and set $K = G_\varepsilon$ in the notation which follows Definition 3.7; by hypothesis K is a compact subset of G. Now take $s \in G \setminus K$. Then $G_{\varepsilon,s}$ is compact, and so there is a function $f_s \in C_{00}(G)$ such that $f_s(G_{\varepsilon,s}) = \{1\}$ and $f_s(G) \subset \mathbb{I}$. Define

$$g_s = (1 - f_s)(\lambda \cdot \delta_s)\,.$$

Then $g_s \in \mathcal{X}_\omega$ because $\mathcal{X}_\omega$ is translation-invariant and $f_s \in C_{00}(G)$. Also,

$$\begin{aligned} \|g_s\| &= \sup\left\{\frac{|g_s(t)|}{\omega(t)} : t \in G\right\} \\ &\leq \sup\left\{\frac{|\lambda(st)|}{\omega(st)} \cdot \frac{\omega(st)}{\omega(t)} : t \in G \setminus G_{\varepsilon,s}\right\} \\ &\leq \varepsilon\omega(s) \end{aligned}$$

because $\Omega(s,t) < \varepsilon$ whenever $t \in G \setminus G_{\varepsilon,s}$. Clearly we have

$$g_s - \lambda \cdot \delta_s \in C_{00}(G) \subset \mathcal{E}_\omega\,,$$

and so $\langle \Phi,\, g_s\rangle = \langle \Phi,\, \lambda \cdot \delta_s\rangle$. Hence

$$|(\Phi \cdot \lambda)(s)| = |\langle \Phi,\, \lambda \cdot \delta_s\rangle| = |\langle \Phi,\, g_s\rangle| \leq \varepsilon\omega(s)\,.$$

It follows that

$$\{s \in G : |(\Phi \cdot \lambda)(s)|\,/\omega(s) \geq \varepsilon\} \subset K\,,$$

and so $\Phi \cdot \lambda \in \mathcal{E}_\omega$. Similarly, $\lambda \cdot \Phi \in \mathcal{E}_\omega$. □

We can now give our analogue of Theorem 8.19. There is an important difference: the previous condition '$R_\omega^{\Box 2} = \{0\}$' now becomes '$\mathcal{R}_\omega^{\Box 3} = \{0\}$'. We shall see in Theorem 12.12 that usually it is no longer the case that $\mathcal{R}_\omega^{\Box 2} = \{0\}$.

THEOREM 12.9. *Let ω be a weight function on a locally compact group G, and suppose that $\omega(s) \geq 1$ $(s \in G)$ and that $\mathcal{M}_\omega$ is semisimple. Suppose that Ω 0-clusters strongly on $G \times G$. Then:*

(i) $\mathcal{R}_\omega^{\Box} = \ker \Pi_\omega = \mathcal{E}_\omega^\circ$;

(ii) $\mathcal{R}_\omega^{\Box 2} \subset \overline{\mathcal{R}_\omega^{\Box 2}} \subset \mathcal{X}_\omega^\circ \subset \mathcal{E}_\omega^\circ$;

(iii) $\mathcal{R}_\omega^{\Box 3} = \{0\}$;

(iv) $(\mathcal{B}_\omega,\, \Box\,)$ *has the strong Wedderburn decomposition*

$$(\mathcal{B}_\omega,\, \Box\,) = \kappa_\omega(\mathcal{M}_\omega) \ltimes \mathcal{R}_\omega^{\Box}\,.$$

PROOF. By Proposition 12.7, we have $\mathcal{R}_\omega^\Box \subset \ker \Pi_\omega = \mathcal{E}_\omega^\circ$.

Let $\Phi, \Psi \in \mathcal{E}_\omega^\circ$, and take $\lambda \in \mathcal{X}_\omega$. By Lemma 12.8, $\Psi \cdot \lambda \in \mathcal{E}_\omega$, and so $\langle \Phi \Box \Psi, \lambda \rangle = 0$. Thus $\Phi \Box \Psi \in \mathcal{X}_\omega^\circ$. This shows that $\overline{(\mathcal{E}_\omega^\circ)^{\Box 2}} \subset \mathcal{X}_\omega^\circ$.

By Proposition 12.7, $\mathcal{X}_\omega^\circ$ is a left-annihilator ideal of $(\mathcal{B}_\omega, \Box)$, and so $\mathcal{B}_\omega \Box (\mathcal{E}_\omega^\circ)^{\Box 2} = \{0\}$. In particular $(\mathcal{E}_\omega^\circ)^{\Box 3} = \{0\}$, and so $\mathcal{E}_\omega^\circ$ is a nilpotent ideal in $(\mathcal{B}_\omega, \Box)$. Thus $\mathcal{E}_\omega^\circ \subset \mathcal{R}_\omega^\Box$, and this gives (i), and hence (ii) and (iii). Clause (iv) now follows from (7.25). $\Box$

We now investigate when equality can occur in each of the inclusions of clause (ii) of the above theorem.

THEOREM 12.10. *Let ω be a weight function on a locally compact group G, and suppose that $\omega(s) \geq 1$ $(s \in G)$ and that $\mathcal{M}_\omega$ is semi-simple. Suppose that Ω 0-clusters strongly on $G \times G$, and suppose that G is not discrete. Then $\overline{\mathcal{R}_\omega^{\Box 2}}$ has infinite codimension in $\mathcal{X}_\omega^\circ$.*

PROOF. It is sufficient to find a sequence (λ_n) in $\mathcal{A}'_\omega$ such that $\langle \Phi \Box \Psi, \lambda_n \rangle = 0$ whenever $\Phi, \Psi \in \mathcal{E}_\omega^\circ$ and $n \in \mathbb{N}$ and such that the set $\{\lambda_n + \mathcal{X}_\omega : n \in \mathbb{N}\}$ is linearly independent in $\mathcal{A}'_\omega / \mathcal{X}_\omega$.

The group G contains a pairwise disjoint sequence (K_n) of compact subsets of G such that each set K_n is not open. Set $\lambda_n = \chi_{K_n}$ $(n \in \mathbb{N})$. Then $\{\lambda_n + \mathcal{X}_\omega : n \in \mathbb{N}\}$ is linearly independent in $\mathcal{A}'_\omega / \mathcal{X}_\omega$.

Let $n \in \mathbb{N}$, and take $\Phi, \Psi \in \mathcal{E}_\omega^\circ$. We have

$$\langle f, \Psi \cdot \lambda_n \rangle = \langle \Psi, \lambda_n \cdot f \rangle = 0 \quad (f \in \mathcal{A}_\omega)$$

by Proposition 7.17(iv), and so $\Psi \cdot \lambda_n = 0$. Hence $\langle \Phi \Box \Psi, \lambda_n \rangle = 0$.

Thus (λ_n) has the required properties. $\Box$

We now wish to show that usually $\mathcal{R}_\omega^{\Box 2}$ is infinite-dimensional; we shall require some mild extra conditions on G and ω for our proof. A function λ on G *has period* t (where $t \in G$) if $\lambda(st) = \lambda(s)$ $(s \in G)$.

LEMMA 12.11. *Let ω be a weight function on a locally compact group G such that ω is almost left-invariant, and let $\lambda \in L^\infty(G)$ have period t for some $t \in G$ such that $\operatorname{Lim}_{n\to\infty} t^n = \infty$. Then*

$$\lim_{n\to\infty} (\lambda\omega \cdot f)(t^n)/\omega(t^n) = \langle f, \lambda \rangle \quad (f \in \mathcal{A}_\omega).$$

PROOF. For each $n \in \mathbb{N}$, define

$$F_n(s) = \lambda(st^n)\omega(st^n)/\omega(t^n) \quad (s \in G).$$

We note that

$$\lim_{n\to\infty} |F_n(s) - \lambda(s)| = \lim_{n\to\infty} |\lambda(s)| \left| \frac{\omega(st^n)}{\omega(t^n)} - 1 \right| = 0 \quad (s \in G)$$

because ω is almost left-invariant and $\operatorname{Lim}_{n\to\infty} t^n = \infty$. Let $f \in \mathcal{A}_\omega$. Then we have $|f(s)F_n(s)| \leq \|\lambda\|_\infty |f(s)|\,\omega(s)$ $(s \in G,\ n \in \mathbb{N})$ and

$$\int_G |f(s)|\,\omega(s)\,\mathrm{dm}(s) = \|f\|_\omega < \infty\,,$$

and so, by the dominated convergence theorem,

$$\frac{(\lambda\omega \cdot f)(t^n)}{\omega(t^n)} = \int_G f(s)F_n(s)\,\mathrm{d}m(s) \to \int_G f(s)\lambda(s)\,\mathrm{d}m(s) = \langle f, \lambda\rangle\,.$$

This is the required result. □

We now come to the 'mild extra condition' on the locally compact group G. We require that G contains a compact set K and an element t such that:

(i) K is a symmetric neighbourhood of e_G, and K is not open;

(ii) $\operatorname{Lim}_{n\to\infty} t^n = \infty$ and $t^mK \cap t^nK = \emptyset$ whenever $m, n \in \mathbb{Z}$ with $m \neq n$.

Set $U = \operatorname{int} K$, and take $s_0 \in K \cap \overline{(G \setminus K)}$, the frontier of K in G.

We *claim* that there is now a sequence (s_m) in U such that:

(iii) $s_1 = e_G$ and $s_{m+1}s_0 \in s_1U \cap \cdots \cap s_mU$ for each $m \in \mathbb{N}$.

Indeed take a sequence (r_j) in U such that $r_1 = e_G$ and $r_j \to s_0^{-1}$. We have $e_G \in r_jU$ for each $j \in \mathbb{N}$. Set $s_1 = r_1$. Since $r_js_0 \to e_G$, we can inductively choose a subsequence (s_m) of (r_j) such that

$$s_{m+1}s_0 \in s_1U \cap \cdots \cap s_mU$$

for each $m \in \mathbb{N}$. This sequence (s_m) satisfies (iii).

Let G satisfy the above conditions. For $n \in \mathbb{Z}$, we take χ_n to be the characteristic function of t^nK, and then, for each $m \in \mathbb{N}$, we define

$$\gamma_m(s) = \sum_{n\in\mathbb{Z}} \chi_n(s_m^{-1}s) \quad (s \in G)\,.$$

Clearly each function γ_m belongs to $L^\infty(G)$ with $\|\gamma_m\|_\infty = 1$, and γ_m has period t. Further, we have $\gamma_m(s_ms_0) = 1$, whilst γ_m takes the value 0 in each neighbourhood of s_ms_0; on the other hand, for each $m \geq 2$, there is a neighbourhood U of s_ms_0 such that each of the functions $\gamma_1, \ldots, \gamma_{m-1}$ is continuous on U.

Let ω be a weight function on G, and define

$$H = \left\{\lambda \in \mathcal{X}_\omega : \lim_{n\to\infty} \lambda(t^n)/\omega(t^n) \text{ exists}\right\}.$$

Clearly H is a closed linear subspace of $\mathcal{X}_\omega \subset \mathcal{A}'_\omega$ and $\mathcal{E}_\omega \subset H$. Set

$$\langle \Lambda, \lambda\rangle = \lim_{n\to\infty} \lambda(t^n)/\omega(t^n) \quad (\lambda \in H)\,.$$

Then $\Lambda \in H'$ with $\|\Lambda\| = 1$; clearly $\Lambda \mid \mathcal{E}_\omega = 0$.

Next, for each $m \in \mathbb{N}$, define

$$H_m = H + \operatorname{lin}\{\gamma_1, \ldots, \gamma_m\}.$$

Then H_m is also a closed linear subspace of $\mathcal{A}'_\omega$, and there is a constant $c > 0$ such that $\|\gamma_{m+1} - \lambda\| \geq c$ for each $\lambda \in H_m$. It follows that, for each $m \in \mathbb{N}$, there exists $\Phi_m \in \mathcal{B}_\omega$ such that

$$\langle \Phi_m, \lambda \rangle = \langle \Lambda, \lambda \rangle \quad (\lambda \in H), \quad \langle \Phi_m, \gamma_j \rangle = 0 \quad (j \in \mathbb{N}_{m-1}),$$

and $\langle \Phi_m, \gamma_m \rangle = 1$. We have $\Phi_m \in \mathcal{E}_\omega^\circ$.

A version of the following theorem (in the case where $G = \mathbb{R}$) is given in [La, Proposition 2.2.22].

THEOREM 12.12. *Let G be a locally compact group satisfying the above conditions, and let ω be an almost left-invariant weight on G. Then $\mathcal{R}_\omega^{\square 2}$ is infinite-dimensional.*

PROOF. We use the notations given above.

Fix $m \in \mathbb{N}$. For each $j \in \mathbb{N}_m$, the function $\gamma_j\omega$ belongs to $\mathcal{A}'_\omega$. Let $f \in \mathcal{A}_\omega$. Then $\gamma_j\omega \cdot f$ belongs to $\mathcal{X}_\omega$ by Proposition 7.17(i), and

$$\lim_{n\to\infty} (\gamma_j\omega \cdot f)(t^n)/\omega(t^n) = \langle f, \gamma_j \rangle$$

by Lemma 12.11. Thus $\gamma_j\omega \cdot f \in H$, and

$$\langle f, \Phi_m \cdot \gamma_j\omega \rangle = \langle \Phi_m, \gamma_j\omega \cdot f \rangle = \langle \Lambda, \gamma_j\omega \cdot f \rangle = \langle f, \gamma_j \rangle.$$

It follows that $\Phi_m \cdot \gamma_j\omega = \gamma_j \ (j \in \mathbb{N})$, and hence that

$$\langle \Phi_m \,\square\, \Phi_m, \gamma_j\omega \rangle = \langle \Phi_m, \gamma_j \rangle = \delta_{j,m} \quad (j \in \mathbb{N}_m).$$

This implies that the set $\{\Phi_1 \,\square\, \Phi_1, \ldots, \Phi_m \,\square\, \Phi_m\}$ is linearly independent in $\mathcal{R}_\omega^{\square 2}$.

The result follows. $\square$

For example, set $G = \mathbb{R}$, and consider ω_α, where $\alpha > 0$. Then G and ω_α satisfy all the conditions in both Theorem 12.9 and Theorem 12.12, and so $\mathcal{R}_{\omega_\alpha} = \mathcal{E}_{\omega_\alpha}^\circ$ and $\mathcal{R}_{\omega_\alpha}^{\square 2}$ are each infinite-dimensional.

THEOREM 12.13. *Let G be a σ-compact, locally compact group which is neither discrete nor compact, and let ω be a weight function on G. Then $\mathcal{X}_\omega^\circ$ has infinite codimension in $\mathcal{E}_\omega^\circ$.*

PROOF. Let K be a compact neighbourhood of e_G. Then there is a sequence (s_m) of distinct points contained in $\operatorname{int} K$ such that the set $\{s_m : m \in \mathbb{N}\}$ is discrete in the relative topology. For each $j \in \mathbb{N}$, take $f_j \in C(G)$ such that $\operatorname{supp} f_j \subset \operatorname{int} K$ and $f_j(s_n) = \delta_{j,n} \ (n \in \mathbb{N})$.

Since the space G is σ-compact and not compact, there is a sequence (t_n) in G such that $\operatorname{Lim}_{n\to\infty} t_n = \infty$; we may suppose that $t_1 = e_G$ and that $t_m K \cap t_n K = \emptyset$ whenever $m, n \in \mathbb{N}$ with $m \neq n$. For each $j \in \mathbb{N}$, define

$$\lambda_j(s) = f_j(st_n^{-1}) \quad (s \in G)\,.$$

Then $\lambda_j \in LUC(G)$ and $\lambda_j\omega \in \mathcal{X}_\omega$. Note that

$$\lambda_j(s_m t_n) = \delta_{j,m} \quad (j, m, n \in \mathbb{N})\,.$$

Now fix $m \in \mathbb{N}$, and define

$$H_m = \{\lambda \in CB(G, 1/\omega) : \lim_{n\to\infty} \lambda(s_m t_n)/\omega(s_m t_n) \quad \text{exists}\}\,.$$

Clearly H_m is a closed linear subspace of $\mathcal{A}'_\omega$ and $\mathcal{E}_\omega \subset H_m$. Set

$$\langle \Lambda_m,\, \lambda\rangle = \lim_{n\to\infty} \lambda(s_m t_n)/\omega(s_m t_n) \quad (\lambda \in H_m)\,.$$

Then $\Lambda_m \in H'_m$ with $\|\Lambda_m\| = 1$; clearly $\Lambda_m \mid \mathcal{E}_\omega = 0$. There exists $\Phi_m \in \mathcal{B}_\omega$ such that $\Phi_m \mid H_m = \Lambda_m$, and certainly $\Phi_m \in \mathcal{E}_\omega^\circ$.

For each $j, m \in \mathbb{N}$, we have $\langle \Phi_m,\, \lambda_j\omega\rangle = \delta_{j,m}$, and so it follows that the set $\{\Phi_m + \mathcal{X}_\omega^\circ : m \in \mathbb{N}\}$ is linearly independent in the space $\mathcal{E}_\omega^\circ/\mathcal{X}_\omega^\circ$.

The result follows. □

We summarize our recent results in the following theorem, which applies in the special case where the group G is the real line $\mathbb{R}$. The result is essentially [La, Theorem 2.2.29].

THEOREM 12.14. *Let ω be an almost invariant weight function on $\mathbb{R}$ such that $1/\omega \in C_0(\mathbb{R})$. Then $(\mathcal{B}_\omega, \Box\,) = \mathcal{M}_\omega \ltimes \mathcal{E}_\omega^\circ$ and*

$$\{0\} = \mathcal{R}_\omega^{\Box 3} \subsetneq \mathcal{R}_\omega^{\Box 2} \subset \overline{\mathcal{R}_\omega^{\Box 2}} \subsetneq \mathcal{X}_\omega^\circ \subsetneq \mathcal{E}_\omega^\circ = \mathcal{R}_\omega\,.$$

Further:

(i) *$\mathcal{R}_\omega^{\Box 2}$ is infinite-dimensional;*

(ii) *$\overline{\mathcal{R}_\omega^{\Box 2}}$ has infinite codimension in $\mathcal{X}_\omega^\circ$;*

(iii) *$\mathcal{X}_\omega^\circ$ has infinite codimension in $\mathcal{E}_\omega^\circ$.*

PROOF. This follows from equation (7.25) and Theorems 12.9, 12.10, 12.12, and 12.13. □

We do not know whether or not the space $\mathcal{R}_\omega^{\Box 2}$ in the above theorem is necessarily closed.

CHAPTER 13

Derivations into Second Duals

Our aim in the present chapter is to investigate when the Beurling algebras $\ell^1(G,\omega)$ and $L^1(G,\omega)$ are 2-weakly amenable (see Definition 1.5). Our techniques apply only to commutative algebras, and so throughout this section we shall consider abelian groups G (and we shall write the group operation additively). The theorems proved here are developments of those in [La, Chapter 3].

We shall prove, for example, that the commutative Banach algebras $\ell^1(\omega_\alpha) = \ell^1(\mathbb{Z},\omega_\alpha)$ and $L^1(\mathbb{R},\omega_\alpha)$ are 2-weakly amenable if and only if $\alpha < 1$.

We shall use the following basic formula. Let A be a commutative, unital algebra, let E be a unital A-module, let $D : A \to E$ be a derivation, and let $a \in \operatorname{Inv} A$. Then

$$(13.1) \qquad D(a^n) = na^{n-1} \cdot Da \quad (n \in \mathbb{Z}),$$

and so

$$(13.2) \qquad Da = \frac{1}{n}a^{1-n} \cdot D(a^n) \quad (n \in \mathbb{Z}).$$

We begin with the case where ω is a weight on a group G. We continue to use the notations $A_\omega = \ell^1(G,\omega)$, A'_ω, E_ω, and B_ω specified in (8.1).

THEOREM 13.1. *Let G be an abelian group, and let ω be a weight on G such that ω is almost invariant and* $\inf_{n\in\mathbb{N}} \omega(nt)/n = 0$ *for each $t \in G$. Then A_ω is 2-weakly amenable.*

PROOF. Let $D : A_\omega \to B_\omega$ be a continuous derivation; we may suppose that $\|D\| \le 1$.

Let $\pi_\omega : B_\omega \to A_\omega$ be the canonical projection, as in (8.2). Then the map $\pi_\omega \circ D : A_\omega \to A_\omega$ is a continuous derivation; also A_ω is a semisimple Banach algebra, and so, by the Singer–Wermer theorem [D, Theorem 2.7.20], $\pi_\omega \circ D = 0$. Thus $D(A_\omega) \subset \ker \pi_\omega = E_\omega^\circ$.

To prove that $D = 0$, it suffices to show that $\langle D(\delta_t), \lambda\rangle = 0$ for each $t \in G$ and $\lambda \in A'_\omega$. Fix $t \in G$ and $\lambda \in A'_\omega$.

Let $n \in \mathbb{N}$. We define

$$H_n = \{s \in G : \omega(s + (1-n)t) \geq 2\omega(s)\}\,.$$

Since ω is almost invariant, the set H_n is finite. Define

$$\sigma_n(s) = \begin{cases} 0 & (s \in H_n)\,, \\ (\delta_{(1-n)t} \cdot \lambda)(s) & (s \in G \setminus H_n)\,. \end{cases}$$

We have

$$\begin{aligned} \sup_{s\in G} \frac{|\sigma_n(s)|}{\omega(s)} &= \sup_{s \in G\setminus H_n} \left\{ \frac{|\lambda(s+(1-n)t)|}{\omega(s+(1-n)t)} \cdot \frac{\omega(s+(1-n)t)}{\omega(s)} \right\} \\ &\leq 2\left\|\lambda\right\|\,, \end{aligned}$$

and so $\sigma_n \in A'_\omega$ with $\|\sigma_n\| \leq 2\|\lambda\|$. Since $\sigma_n - \delta_{(1-n)t} \cdot \lambda \in E_\omega$ and $D(\delta_{nt}) \in E^\circ_\omega$, we have

$$\langle \delta_{(1-n)t} \cdot D(\delta_{nt}),\, \lambda\rangle = \langle D(\delta_{nt}),\, \delta_{(1-n)t} \cdot \lambda\rangle = \langle D(\delta_{nt}),\, \sigma_n\rangle,$$

and so it follows from (13.2) that

$$|\langle D(\delta_t),\, \lambda\rangle| \leq \frac{1}{n}\,|\langle D(\delta_{nt}),\, \sigma_n\rangle| \leq 2\left\|\lambda\right\|\omega(nt)/n\,.$$

Since $\inf_{n\in\mathbb{N}} \omega(nt)/n = 0$, we see that $\langle D(\delta_t),\, \lambda\rangle = 0$. Thus $D = 0$, as required. □

Let ω be a weight on $\mathbb{Z}$. Then the method of the above proof shows that $\ell^1(\omega)$ is 2-weakly amenable whenever the following condition on ω is satisfied: there is an infinite subset S of $\mathbb{Z}^+$ and a function

$$n \mapsto \alpha_n, \quad S \to \mathbb{R}^+ \setminus \{0\}\,,$$

such that $\inf_{n\in S} \alpha_n\omega(n)/n = 0$ and the set

$$\{k \in \mathbb{Z} : \omega(k+n-1) \geq \alpha_n\omega(k)\}$$

is finite for each $n \in S$.

For example, let $\widetilde{\omega}_\alpha = \omega_\alpha\omega$, where ω is the weight specified in Example 9.17. Then $\widetilde{\omega}_\alpha$ satisfies the above conditions whenever $\alpha < 1$ (taking

$$S = \{2^j + 1 : j \in \mathbb{N}\})\,,$$

and so $\ell^1(\widetilde{\omega}_\alpha)$ is 2-weakly amenable whenever $\alpha < 1$, even though $\widetilde{\omega}_\alpha$ is not almost invariant. Again, let ω be a weight specified in Example 9.12, with $\varphi(n) = \log(1+n)$ $(n \in \mathbb{Z}^+)$. Then $\ell^1(\omega_\alpha\omega)$ is 2-weakly amenable whenever $\alpha < 1$.

In fact, we *conjecture* that $\ell^1(G,\omega)$ is 2-weakly amenable whenever ω is a weight on a group G such that $\inf_{n\in\mathbb{N}} \omega(nt)/n = 0$ $(t \in G)$.

The following result shows that the condition on the rate of growth of $\omega(nt)$ is necessary.

THEOREM 13.2. *Let $\alpha \in \mathbb{R}^+$. Then the Beurling algebra $\ell^1(\omega_\alpha)$ is 2-weakly amenable if and only if $\alpha < 1$.*

PROOF. First suppose that $\alpha < 1$. Then the weight ω_α on $\mathbb{Z}$ satisfies the conditions on ω which were specified in Theorem 13.1, and so $\ell^1(\omega_\alpha)$ is 2-weakly amenable.

Now suppose that $\alpha \geq 1$, so that the Fourier transform $\widehat{f}$ of a function $f \in \ell^1(\omega_\alpha)$ is a continuously differentiable function on $\mathbb{T}$. The weight ω_α is almost invariant, and so, by Theorem 7.38, there is an invariant mean, say M, on $\ell^1(\omega_\alpha)'$, where we note that $\mathcal{L}_t(\mathbb{R}) \neq \emptyset$. Let $f = \sum_{n\in\mathbb{Z}} \alpha_n\delta_n \in \ell^1(\omega_\alpha)$. Then

$$f \cdot \mathrm{M} = \sum_{n\in\mathbb{Z}} \alpha_n\delta_n \cdot \mathrm{M} = \left(\sum_{n\in\mathbb{Z}} \alpha_n\right)\mathrm{M} = \widehat{f}(1)\mathrm{M}$$

because $\delta_n \cdot \mathrm{M} = \mathrm{M}$ for each $n \in \mathbb{Z}$. Define

$$D : f \mapsto \widehat{f}\,'(1)\mathrm{M}, \quad \ell^1(\omega_\alpha) \to \ell^1(\omega_\alpha)''\,.$$

Clearly D is a non-zero, continuous linear map. Let $f, g \in \ell^1(\omega_\alpha)$. Then we have

$$\begin{aligned} D(f \star g) &= (\widehat{f \star g})'(1)\mathrm{M} = \widehat{f}\,'(1)\widehat{g}(1)\mathrm{M} + \widehat{f}(1)\widehat{g}\,'(1)\mathrm{M} \\ &= \widehat{f}\,'(1)g \cdot \mathrm{M} + \widehat{g}\,'(1)f \cdot \mathrm{M} = D(f) \cdot g + f \cdot Dg\,, \end{aligned}$$

and so D is a derivation.

Thus $\ell^1(\omega_\alpha)$ is not 2-weakly amenable. □

It might be thought, on consideration of the above theorem, that there will be a non-zero, continuous derivation from $\ell^1(\omega)$ into $\ell^1(\omega)''$ whenever $\{\widehat{f} : f \in \ell^1(\omega)\}$ consists of continuously differentiable functions on $\mathbb{T}$. However, the next result shows that this is not the case.

THEOREM 13.3. *Let $\omega(n) = \exp(|n|)$ $(n \in \mathbb{Z})$. Then $\widehat{f}$ is analytic on the annulus $\{z \in \mathbb{C} : 1/\mathrm{e} < |z| < \mathrm{e}\}$ for each $f \in \ell^1(\omega)$. However $\ell^1(\omega)$ is 2-weakly amenable.*

PROOF. As we remarked in Chapter 7, $\widehat{f} \in A(X)$ for each function $f \in \ell^1(\omega)$, where we set $X = \{z \in \mathbb{C} : 1/\mathrm{e} \leq |z| \leq \mathrm{e}\}$, and so $\widehat{f}$ is analytic on the specified annulus.

Set $A_\omega = \ell^1(\omega)$, so that

$$B_\omega = A_\omega'' = A_\omega \oplus E_\omega^{\circ +} \oplus E_\omega^{\circ -}\,,$$

as in (8.9).

Let $D : A_\omega \to B_\omega$ be a continuous derivation, and set $D(\delta_1) = \Lambda$. Then $\Lambda \in E_\omega^\circ$ and $\Lambda = \Lambda^+ + \Lambda^-$, where $\Lambda^+ \in E_\omega^{\circ +}$ and $\Lambda^- \in E_\omega^{\circ -}$.

For each $k \in \mathbb{Z}^+$, the map $f \mapsto \delta_k \star f$ is a linear isomorphism on $A_\omega(\mathbb{Z}^+)$ and

$$\|\delta_k \star f\| = \omega(k)\,\|f\| \quad (f \in A_\omega(Z^+))\,.$$

Thus the map $\Phi \mapsto \delta_k \cdot \Phi$ is a linear isomorphism on $B_\omega(\mathbb{Z}^+)$ with

$$\|\delta_k \cdot \Phi\| = \omega(k)\,\|\Phi\| \quad (\Phi \in B_\omega(\mathbb{Z}^+))\,.$$

Now let $n \in \mathbb{N}$. Then it follows from (13.1) that

$$\|D\|\,\omega(n) \geq \|\delta_{n-1} \cdot \Lambda\| \geq \left\|\delta_{n-1} \cdot \Lambda^+\right\| = n\omega(n-1)\left\|\Lambda^+\right\|\,.$$

Thus $\|\Lambda^+\| \leq \mathrm{e}\,\|D\|\,/n$. This is true for each $n \in \mathbb{N}$, and so $\Lambda^+ = 0$. Similarly, $\Lambda^- = 0$, and so $D(\delta_1) = \Lambda = 0$. By the same argument, we have $D(\delta_{-1}) = 0$, and so $D = 0$.

Thus A_ω is 2-weakly amenable. □

We now turn to the case of a general locally compact abelian group G. Let ω be a weight function on G. We again use the notations $\mathcal{A}_\omega = L^1(G,\omega)$, $\mathcal{A}'_\omega$, $\mathcal{B}_\omega$, $\mathcal{M}_\omega$, and $\mathcal{X}_\omega$ from (7.16).

PROPOSITION 13.4. *Let ω be a weight function on a locally compact abelian group G. Let $D : \mathcal{A}_\omega \to \mathcal{B}_\omega$ be a continuous derivation. Then there is a continuous derivation $\widetilde{D} : \mathcal{M}_\omega \to \mathcal{B}_\omega$ such that $\widetilde{D} \mid \mathcal{A}_\omega = D$.*

PROOF. By Theorem 7.14, we can identify the multiplier algebra of $\mathcal{A}_\omega$ with $\mathcal{M}_\omega$. The result is now a special case of Proposition 1.12(ii). □

We next define a locally complex topology on the space $\mathcal{M}_\omega$. For each $f \in \mathcal{A}_\omega$ and $\lambda \in \mathcal{X}_\omega$, define seminorms $\|\cdot\|_f$ and $\|\cdot\|_\lambda$, respectively, by the formulae:

$$\|\mu\|_f = \|\mu \star f\|\,, \quad \|\mu\|_\lambda = \|\mu \cdot \lambda\| \quad (\mu \in \mathcal{M}_\omega)\,.$$

The *τ-topology* on $\mathcal{M}_\omega$ is the topology determined by the family

$$\left\{\|\cdot\|_f\,,\,\|\cdot\|_\lambda : f \in \mathcal{A}_\omega,\, \lambda \in \mathcal{X}_\omega\right\}$$

of seminorms. Clearly $(\mathcal{M}_\omega, \tau)$ is a locally convex space, and τ is stronger than the topology $\|\cdot\|$.

LEMMA 13.5. *Let ω be a weight function on a locally compact abelian group G, and let $\mu \in \mathcal{M}_\omega$. Then:*

(i) *there is a net (μ_α) in the subspace* $\text{lin}\{\delta_s : s \in G\}$ *such that* $\mu_\alpha \xrightarrow{\tau} \mu$ *and* $\|\mu_\alpha\| \leq \|\mu\|$ *for each* α;

(ii) *there is a net (μ_α) in $\mathcal{A}_\omega$ such that $\mu_\alpha \xrightarrow{\tau} \mu$ and $\|\mu_\alpha\| \leq \|\mu\|$ for each α.*

PROOF. This is similar to a standard proof given in [D, Proposition 3.3.41], and we just sketch the method. Set $L = \text{lin}\{\delta_s : s \in G\}$.

First take χ_K to be the characteristic function of a compact, symmetric neighbourhood K of e_G in G. A routine calculation shows that there is an element ν of L with $\|\nu\| \leq \|\chi_K\|$ in each prescribed τ-neighbourhood of χ_K, and so $\chi_K \in \overline{L}^{(\tau)}$.

The space $\overline{L}^{(\tau)} \cap \mathcal{A}_\omega$ is a translation-invariant, $\|\cdot\|$-closed linear subspace of $\mathcal{A}_\omega$, and so it is a closed ideal in $\mathcal{A}_\omega$. Since $\mathcal{A}_\omega$ has a bounded approximate identity, say (e_α), the space $\overline{L}^{(\tau)} \cap \mathcal{A}_\omega$ is a closed ideal in $\mathcal{M}_\omega$, and this ideal contains (e_α), and hence the net $(\mu \star e_\alpha)$.

For each $f \in \mathcal{A}_\omega$ and each α, we have

$$\|(\mu \star e_\alpha) \star f - \mu \star f\| \leq \|\mu\| \, \|e_\alpha \star f - f\| \, ,$$

and, for each $\lambda \in \mathcal{X}_\omega$ and each α, we have

$$\|(\mu \star e_\alpha) \cdot \lambda - \mu \cdot \lambda\| \leq \|\mu\| \, \|e_\alpha \cdot \lambda - \lambda\| \, .$$

Since (e_α) is also a bounded approximate identity for the essential module $\mathcal{X}_\omega$, it follows that $\tau - \lim_\alpha \mu \star e_\alpha = \mu$.

The result follows. □

LEMMA 13.6. *Let ω be a weight function on a locally compact abelian group G. Let $D : \mathcal{M}_\omega \to \mathcal{B}_\omega$ be a continuous derivation. Suppose that $\mu \in \mathcal{M}_\omega$ and that (μ_α) is a bounded net in $\mathcal{M}_\omega$ with $\tau - \lim_\alpha \mu_\alpha = \mu$. Then*

$$\lim_\alpha \langle D(\mu_\alpha), \lambda \rangle = \langle D\mu, \lambda \rangle \quad (\lambda \in \mathcal{X}_\omega) \, .$$

PROOF. We may suppose that $\|\mu\| \leq 1$ and that there is a constant $C > 0$ such that $\|\mu_\alpha\| \leq C$ for each α. We may also suppose that $\|D\| \leq 1$.

Take $\lambda \in \mathcal{X}_\omega$ with $\|\lambda\| \leq 1$, and take $\varepsilon > 0$. Since $\mathcal{A}_\omega$ has an approximate identity in $(\mathcal{A}_\omega)_{[1]}$ and $\mathcal{X}_\omega$ is an essential module, there exists $f \in \mathcal{A}_\omega$ with

$$\|f \cdot \lambda - \lambda\| < \varepsilon \quad \text{and} \quad \|f\| \leq 1 \, .$$

Since $\tau - \lim_\alpha \mu_\alpha = \mu$, there exists α_0 such that

$$\|f \star \mu_\alpha - f \star \mu\| < \varepsilon \quad \text{and} \quad \|\mu_\alpha \cdot \lambda - \mu \cdot \lambda\| < \varepsilon$$

whenever $\alpha \succeq \alpha_0$. Set $\nu_\alpha = \mu_\alpha - \mu$, so that $\|\nu_\alpha\| \leq C + 1$ for each α. For each $\alpha \succeq \alpha_0$, we have

$$\begin{aligned} |\langle D(\nu_\alpha), \lambda\rangle| &\leq |D(\nu_\alpha), \lambda - f \cdot \lambda\rangle| + |\langle D(\nu_\alpha) \star f, \lambda\rangle| \\ &\leq (C+1)\varepsilon + |\langle D(\nu_\alpha \star f), \lambda\rangle| + |\langle Df, \nu_\alpha \cdot \lambda\rangle| \\ &\leq (C+1)\varepsilon + \varepsilon + \varepsilon = (C+3)\varepsilon\,. \end{aligned}$$

Thus $\lim_\alpha \langle D\nu_\alpha, \lambda\rangle = 0$, as required. □

THEOREM 13.7. *Let G be a locally compact abelian group, let ω be an almost invariant weight function on G such that $\omega(s) \geq 1$ $(s \in G)$ and $\omega(nt) = o(n)$ as $n \to \infty$ for each $t \in G$, and let $D : \mathcal{M}_\omega \to \mathcal{B}_\omega$ be a continuous derivation. Then:*

(i) *$D(\delta_t) = 0 \quad (t \in G)$;*

(ii) *$D(\mu) \in \mathcal{X}_\omega^\circ \quad (\mu \in \mathcal{M}_\omega)$.*

PROOF. As in (7.25), we have the decomposition $\mathcal{B}_\omega = \mathcal{M}_\omega \ltimes \mathcal{E}_\omega^\circ$, and the canonical projection $\Pi_\omega : \mathcal{B}_\omega \to \mathcal{M}_\omega$ is a continuous epimorphism. Since the commutative Banach algebra $\mathcal{M}_\omega$ is semisimple, it again follows from the Singer–Wermer theorem that $D(\mathcal{A}_\omega) \subset \mathcal{E}_\omega^\circ$. We may suppose that $\|D\| \leq 1$.

To prove that $D(\delta_t) = 0$ $(t \in G)$, we follow the argument given in the proof of Theorem 13.1 with $\lambda \in \mathcal{A}_\omega'$; now, for each $n \in \mathbb{N}$, H_n is a compact subset of G and $\sigma_n \in \mathcal{A}_\omega'$ with $\|\sigma_n\| \leq 2\,\|\lambda\|$ because ω is almost invariant. As before, $\langle D(\delta_t), \lambda\rangle = 0$ for each $\lambda \in \mathcal{A}_\omega'$, and so $D(\delta_t) = 0$. This gives (i).

Let $\mu \in \mathcal{M}_\omega$. By Lemma 13.5(i), there is a net (μ_α) in the subspace $\mathrm{lin}\{\delta_s : s \in G\}$ such that $\mu_\alpha \xrightarrow{\tau} \mu$ and $\|\mu_\alpha\| \leq \|\mu\|$ for each α. We have shown that $D(\mu_\alpha) = 0$ for each α. By Lemma 13.6,

$$\langle D\mu, \lambda\rangle = 0 \;\; (\lambda \in \mathcal{X}_\omega)\,,$$

and so $D\mu \in \mathcal{X}_\omega^\circ$. This gives (ii). □

THEOREM 13.8. *Let G be a locally compact abelian group, and let ω be an almost invariant weight function on G such that $\omega(s) \geq 1$ $(s \in G)$ and $\omega(nt) = o(n)$ as $n \to \infty$ for each $t \in G$. Then $L^1(G, \omega)$ is 2-weakly amenable.*

PROOF. The weight function ω on G satisfies the conditions in Theorem 13.7.

Let $D : \mathcal{A}_\omega \to \mathcal{B}_\omega$ be a continuous derivation. By Proposition 13.4, there is a continuous derivation $\widetilde{D} : \mathcal{M}_\omega \to \mathcal{B}_\omega$ such that $\widetilde{D} \mid \mathcal{A}_\omega = D$. By Theorem 13.7(ii), $\widetilde{D}(\mu) \in \mathcal{X}_\omega^\circ$ $(\mu \in \mathcal{M}_\omega)$; in particular, we have

$$D(f) \in \mathcal{X}_\omega^\circ \ (f \in \mathcal{A}_\omega)\,.$$

Now take $f \in \mathcal{A}_\omega$. Since $\mathcal{A}_\omega$ has a bounded approximate identity, there exist $g, h \in \mathcal{A}_\omega$ such that $f = g \star h$, and then

$$Df = g \cdot Dh + Dg \cdot h \in \mathcal{A}_\omega \cdot \mathcal{X}_\omega^\circ\,.$$

However, by Proposition 12.7, $\mathcal{X}_\omega^\circ$ is a left-annihilator ideal of $\mathcal{B}_\omega$, and so $\mathcal{A}_\omega \cdot \mathcal{X}_\omega^\circ = \{0\}$. Thus $D = 0$, and so $\mathcal{A}_\omega$ is 2-weakly amenable. □

THEOREM 13.9. *Let $\alpha \in \mathbb{R}^+$. Then the Beurling algebra $L^1(\mathbb{R}, \omega_\alpha)$ is 2-weakly amenable if and only if $\alpha < 1$.*

PROOF. First suppose that $\alpha < 1$. Then ω_α satisfies the conditions on ω in Theorem 13.8, and so $L^1(\mathbb{R}, \omega_\alpha)$ is 2-weakly amenable.

Now suppose that $\alpha \geq 1$. Again by Theorem 7.38, there is an invariant mean, say M, on $\mathcal{A}'_\omega$. For each $f \in \mathcal{A}_\omega$, the Fourier transform $\widehat{f}$ of f is a continuously differentiable function on $\mathbb{R}$: define

$$D : f \mapsto \widehat{f}\,'(0)\mathrm{M}, \quad \mathcal{A}_{\omega_\alpha} \to \mathcal{B}_{\omega_\alpha}\,.$$

Then it is clear that D is a non-zero, continuous derivation, and so $\mathcal{A}_{\omega_\alpha}$ is not 2-weakly amenable. □

CHAPTER 14

Open Questions

1. Let E be a Banach space. Is $(\mathcal{B}(E)'', \Box)$ semisimple for sufficiently 'nice' Banach spaces E? (This is true when E is a Hilbert space.) In particular, is $(\mathcal{B}(\ell^p(\mathbb{N}))'', \Box)$ semisimple whenever $1 < p < \infty$?

[*Added in August 2004:* This question has now been essentially resolved by Daws and Read [DaRe].]

2. Let ω be a weight function on a locally compact group G. Is $L^1(G, \omega)$ always semisimple? In particular, is this true whenever G is a discrete group?

3. Let G be a locally compact group, and let ω be a weight function on G. Set $A_\omega = L^1(G, \omega)$. Can the radicals of $(A''_\omega, \Box)$ and $(A''_\omega, \Diamond)$ be distinct sets? In particular, can this happen in the special case where $\omega = 1$? Are there reasonable conditions on G and ω that imply that $\mathcal{R}_\omega^{\Box 2} = 0$, where $\mathcal{R}_\omega^{\Box}$ is the radical of $(A''_\omega, \Box)$? Is this always true in the special case where $\omega = 1$?

4. Give necessary and sufficient conditions on ω for $\ell^1(\mathbb{Z}, \omega)$ to be strongly Arens irregular. In particular, let ω be a weight on $\mathbb{Z}$ such that

$$\liminf_{n \to \infty} \omega(n) < \infty \quad \text{and} \quad \liminf_{n \to \infty} \omega(-n) < \infty .$$

Does it follow that $\ell^1(\mathbb{Z}, \omega)$ is strongly Arens irregular? Are the algebras of Example 8.12 strongly Arens irregular? Is there a weight function ω such that the Beurling algebra $L^1(G, \omega)$ is left, but not right, strongly Arens irregular?

5. Let ω be the weight in Example 8.15. Is it true that $R_\omega^{\Box 2}$ closed in B_ω? Is $\mathfrak{Z}(B_\omega) = A_\omega \oplus R_\omega^{\Box 2}$?

6. Is there a weight ω on $\mathbb{Z}$ such that $(\ell^1(\mathbb{Z}, \omega)'', \Box)$ is semisimple? Is this the case for Feinstein's example, Example 9.17?

7. Let ω be weight on $\mathbb{Z}$. Suppose that $\Phi, \Psi \in E_\omega^\circ$ and that $\Phi \Box \Psi \neq 0$. Does it follow that $\Phi \notin \mathfrak{Z}(\ell^1(\mathbb{Z}, \omega)'')$? Is it always true that

$$\mathfrak{Z}_t^{(1)}(\ell^1(G, \omega)'') \cap E_\omega^\circ \subset R_\omega^\Box \, ?$$

8. Set $\omega_\alpha(t) = (1 + |t|)^\alpha$ $(t \in \mathbb{R})$, and take $\alpha > 0$. What is a characterization of the centre $\mathfrak{Z}(L^1(\mathbb{R}, \omega_\alpha)'')$?

9. Let ω be an almost invariant weight function on $\mathbb{R}$ with $1/\omega \in C_0(\mathbb{R})$. Is $\mathcal{R}_\omega^{\Box 2}$ necessarily closed in $\mathcal{B}_\omega$?

10. Let G be an abelian group, and let ω be a weight on G such that

$$\inf_{n \in \mathbb{N}} \omega(nt)/n = 0$$

for each $t \in G$. Is A_ω necessarily 2-weakly amenable? The weight ω given in Feinstein's example, Example 9.17, satisfies the specified condition: is this algebra A_ω 2-weakly amenable?

Bibliography

[Ak] C. A. Akemann, The dual space of an operator algebra, *Trans. American Math. Soc.*, 126 (1967), 286–302.

[Ar1] R. Arens, Operations induced in function clases, *Monatsh Math.*, 55 (1951), 1–19.

[Ar2] R. Arens, The adjoint of a bilinear operation, *Proc. American Math. Soc.*, 2 (1951), 839–848.

[Ark] N. Arikan, Arens regularity and reflexivity, *Quarterly J. Math. Oxford* (2), 32 (1981), 383–388.

[BCD] W. G. Bade, P. C. Curtis, Jr., and H. G. Dales, Amenability and weak amenability for Beurling and Lipschitz algebras, *Proc. London Math. Soc.* (3), 55 (1987), 359–377.

[BD] W. G. Bade and H. G. Dales, Continuity of derivations from radical convolution algebras, *Studia Mathematica*, 95 (1989), 59–91.

[BDL] W. G. Bade, H. G. Dales, and Z. A. Lykova, Algebraic and strong splittings of extensions of Banach algebras, *Memoirs American Math. Soc.*, Volume 656, (1999).

[BaR] J. W. Baker and A. Rejali, On the Arens regularity of weighted convolution algebras, *J. London Math. Soc.* (2), 40 (1989), 535–546.

[BaLPy] J. W. Baker, A. T.-M. Lau, and J. Pym, Module homomorphisms and topological centres associated with weakly sequentially complete Banach algebras, *J. Functional Analysis*, 158 (1998), 186–208.

[Bar] B. A. Barnes, A note on separating families of representations, *Proc. American Math. Soc.*, 87 (1983), 95–98.

[BJM] J. F. Berglund, H. D. Junghenn, and P. Milnes, *Analysis on semigroups; function spaces, compactifications, representations*, Canadian Math. Soc. Series of Monographs and Advanced Texts, John Wiley, New York, 1989.

[BhDe] S. J. Bhatt and H. V. Dedania, A Beurling algebra is semisimple: an elementary proof, *Bull. Australian Math. Soc.,* 66 (2002), 91–93.

[BoDu] F. F. Bonsall and J. Duncan, *Complete normed algebras*, Springer–Verlag, Berlin, 1973.

[CiY] P. Civin and B. Yood, The second conjugate space of a Banach algebra as an algebra, *Pacific J. Math.*, 11 (1961), 847–870.

[Co] J. B. Conway, *A course in functional analysis*, Springer–Verlag, New York, 1990.

[CrY] I. G. Craw and N. J. Young, Regularity of multiplication in weighted group algebras and semigroup algebras, *Quarterly J. Math. Oxford* (2), 25 (1974), 351–358.

[D] H. G. Dales, *Banach algebras and automatic continuity*, London Math. Society Monographs, Volume 24, Clarendon Press, Oxford, 2000.

[DGhGr] H. G. Dales, F. Ghahramani, and N. Grønbæk, Derivations into iterated duals of Banach algebras, *Studia Mathematica*, 128 (1998), 19–54.

[DGhH] H. G. Dales, F. Ghahramani, and A. Ya. Helemskii, The amenability of measure algebras, *J. London Math. Soc.* (2), 66 (2002), 213–226.

[Da1] M. Daws, Arens regularity of the algebra of operators on a Banach space, *Bull. London Math. Soc.* 36 (2004), 493–503.

[Da2] M. Daws, *Banach algebras of operators*, Thesis, University of Leeds, 2004.

[DaRe] M. Daws and C. J. Read, Semisimplicity of $\mathcal{B}(E)''$, *J. Functional Analysis*, to appear.

[Day] M. M. Day, Amenable semigroups, *Illinois J. Mathematics,* 1 (1957), 509–544.

[DeF] A. Defant and K. Floret, *Tensor norms and operator ideals*, North Holland, Amsterdam, 1993.

[DesGh] M. Despić and F. Ghahramani, Weak amenability of group algebras of locally compact groups, *Canadian Math. Bulletin* (2), 37 (1994), 165–167.

[DiU] J. Diestel and J. J. Uhl, Jr., *Vector measures*, Mathematical Surveys 15, American Mathematical Society, Providence, Rhode Island, 1977.

[Dix] P. G. Dixon, Left approximate identities in algebras of compact operators on Banach spaces, *Proc. Royal Soc. Edinburgh, Section A*, 104 (1986), 169–175.

[DuH] J. Duncan and S. A. R. Hosseiniun, The second dual of a Banach algebra, *Proc. Royal Soc. Edinburgh, Section A*, 84 (1979), 309–325.

[DuU] J. Duncan and A. Ülger, Almost periodic functionals on Banach algebras, *Rocky Mountain J. Math.*, 22 (1992), 837–848.

[DfS] N. Dunford and J. Schwartz, *Linear operators, Part I: General Theory*, Interscience, New York–London, 1958.

[Ed] G. A. Edgar, An ordering for Banach spaces, *Pacific J. Math.* 108 (1983), 83–98.

[ER] E. G. Effros and Z.-J. Ruan, On non-self-adjoint operator algebras, *Proc. American Math. Soc.*, 110 (1990), 915–922.

[El] R. Ellis, Locally compact transformation groups, *Duke Math. J.*, 24 (1957), 119–125.

[FG] G. Fendler, K. Gröchenig, M. Leinert, J. Ludwig, and C. Molitor-Braun, Weighted group algebras of polynomial growth, *Math. Zeitschrift*, 245 (2003), 791–821.

[Fi1] M. Filali, Finite-dimensional right ideals in algebras associated to a locally compact group, *Proc. American Math. Soc.*, 127 (1999), 1729–1734.

[Fi2] M. Filali, Finite-dimensional left ideals in algebras associated to a locally compact group, *Proc. American Math. Soc.*, 127 (1999), 2325–2333.

[FiPy] M. Filali and J. S. Pym, Right cancellation in the LUC-compactifiction of a locally compact group, *Bull. London Math. Soc.*, 35 (2003), 128–134.

[FiSa] M. Filali and and P. Salmi, On one-sided ideals and right cancellation in the second dual of the group algebra and similar algebras, preprint, 2004.

[FiSi] M. Filali and and A. I. Singh, Recent developments on Arens regularity and ideal structures of the second dual of a group algebra, General topological algebras (Tartu, 1999), *Math. Studies (Tartu)*, 1 (2001), 95–124.

[Fo1] B. Forrest, Arens regularity and discrete groups, *Pacific J. Math.*, 151 (1991), 217–227.

[Fo2] B. Forrest, Arens regularity and the $A_p(G)$ algebras, *Proc. American Math. Soc.*, 119 (1993), 595–598.

[GhLaa] F. Ghahramani and J. Laali, Amenability and topological centres of the second duals of Banach algebras, *Bull. Australian Math. Soc.*, 65 (2002), 191–197.

[GhL1] F. Ghahramani and A. T.-M. Lau, Isometric isomorphisms between the second conjugate algebras of group algebras, *Bull. London Math. Soc.*, 20 (1988), 342–344.

[GhL2] F. Ghahramani and A. T.-M. Lau, Multipliers and ideals in the second conjugate algebras related to locally compact groups, *J. Functional Analysis*, 132 (1995), 170–191.

[GhLLos] F. Ghahramani, A. T.-M. Lau, and V. Losert, Isometric isomorphisms between Banach algebras related to locally compact groups, *Trans. American Math. Soc.*, 321 (1990), 273–283.

[GhM] F. Ghahramani and J. P. McClure, Module homomorphisms of the dual modules of convolution Banach algebras, *Canadian Math. Bulletin* (2), 35 (1992), 180–185.

[GhMMe] F. Ghahramani, J. P. McClure, and M. Meng, On the asymmetry of topological centres of the second duals of Banach algebras, *Proc. American Math. Soc.*, 126 (1998), 1765–1768.

[GoT] G. Godefroy and M. Talagrand, Classes d'éspaces de Banach à predual unique, *C. R. Acad. Sciences, Paris*, 292 (1981), 323–325.

[Gou] F. Gourdeau, Amenability and the second dual of a Banach algebra, *Studia Mathematica*, 125 (1997), 75–81.

[Grm1] C. C. Graham, Arens regularity and weak sequential completeness for quotients of the Fourier algebra, *Illinois J. Math.*, 44 (2000), 712–740.

[Grm2] C. C. Graham, Arens regularity and the second dual of certain quotients of the Fourier algebra, *Quarterly J. Math. Oxford* (2), 52 (2001), 13–24.

[Grm3] C. C. Graham, Arens regularity for quotients of the Herz algebra $A_p(E)$, *Bull. London Math. Soc.*, 34 (2002), 457–468.

[Grm4] C. C. Graham, Local existence of $\mathcal{K}$-sets, projective tensor products and Arens regularity for $A(E_1 + \cdots + E_n)$, *Proc. American Math. Soc.*, 132 (2004), 1963–1971.

[GrmMc] C. C. Graham and O. C. McGehee, *Essays in commutative harmonic analysis*, Springer–Verlag, Berlin, Heidelberg, and New York, 1979.

[Gra1] E. E. Granirer, On amenable semigroups with a finite dimensional set of invariant means, I and II, *Illinois J. Maths.*, 7 (1963), 32–48 and 49–58.

[Gra2] E. E. Granirer, Exposed points of convex sets and weak sequential convergence, *Memoirs American Math. Soc.*, Volume 123, (1972).

[Gra3] E. E. Granirer, The radical of $L^\infty(G)^*$, *Proc. American Math. Soc.*, 41 (1973), 321–324.

[Gra4] E. E. Granirer, Day points for quotients of a Fourier algebra $A(G)$, extreme nonergodicity of their duals and extreme non Arens regularity, *Illinois J. Math.*, 40 (1996), 402–419.

[Gra5] E. E. Granirer, On the set of topologically invariant means on an algebra of convolution operators on $L^p(G)$, *Proc. American Math. Soc.*, 124 (1996), 3399–3406.

[Gra6] E. E. Granirer, The Fourier–Herz–Lebesgue Banach algebras, preprint, 2004.

[GraL] E. E. Granirer, and A. T.-M. Lau, Invariant means on locally compact groups, *Illinois J. Math.*, 15 (1971), 249–257.

[Gri] R. I. Grigorchuk, Some results on bounded cohomology. In *Combinatorial and geometric group theory*, London Math. Soc. Lecture Notes, Volume 204, 111–163, Cambridge University Press, 1995.

[Gr1] N. Grønbæk, A characterization of weakly amenable Banach algebras, *Studia Mathematica*, 94 (1989), 149–162.

[Gr2] N. Grønbæk, Constructions preserving weak amenability, *Proc. Centre Math. Analysis, Australian National University*, 21 (1989), 186–202.

[Gr3] N. Grønbæk, Amenability of weighted convolution algebras on locally compact groups, *Trans. American Math. Soc.*, 319 (1990), 765–775.

[G1] M. Grosser, Bidualräume und Vervollständigungen von Banachmoduln, *Lecture Notes in Mathematics*, Volume 717 (1979), Springer–Verlag, Berlin.

[G2] M. Grosser, Arens semi-regularity of the algebra of compact operators, *Illinois J. Math.*, 31 (1987), 544–573.

[GLos] M. Grosser and V. Losert, The norm-strict bidual of a Banach algebra and the dual of $C_u(G)$, *Manuscripta Math.*, 45 (1984), 127–146.

[GMo] S. Grosser and M. Moskowitz, Harmonic analysis on central topological groups, *Trans. American Math. Soc.*, 156 (1971), 419–454.

[Gth] A. Grothendieck, Critères de compacité dans les éspaces fonctionnels généraux, *American J. Mathematics*, 74 (1952), 168–186.

[Gu] S. L. Gulick, Commutativity and ideals in the biduals of topological algebras, *Pacific J. Math.*, 18 (1966), 121–137.

[HaLaus] U. Haagerup and N. J. Laustsen, Weak amenability of C^*-algebras and a theorem of Goldstein. In *Banach algebras '97* (ed. E. Albrecht and M. Mathieu), 223–243, Walter de Gruyter, Berlin, 1998.

[Hei] S. Heinrich, Ultrapowers in Banach space theory, *J. für die Reine und Angewandte Mathematick*, 313 (1980), 72–104.

[He1] A. Ya. Helemskii, *The homology of Banach and topological algebras*, Kluwer Academic Publishers, Dordrecht, 1989.

[He2] A. Ya. Helemskii, *Banach and locally convex algebras*, Clarendon Press, Oxford, 1993.

[HR1] E. Hewitt and K. A. Ross, *Abstract harmonic analysis, Volume I*, (Second edition) Springer–Verlag, Berlin, 1979.

[HR2] E. Hewitt and K. A. Ross, *Abstract harmonic analysis, Volume II*, Springer–Verlag, Berlin, 1970.

[HiSt] N. Hindman and D. Strauss, *Algebra in the Stone-Čech compactification*, de Gruyter, Berlin 1998.

[IPyU] N. Işik, J. Pym, and A. Ülger, The second dual of the group algebra of a compact group, *J. London Math. Soc.* (2), 35 (1987), 135–158.

[J1] B. E. Johnson, Cohomology in Banach algebras, *Memoirs American Math. Soc.*, Volume 127, (1972).

[J2] B. E. Johnson, Weak amenability of group algebras, *Bull. London Math. Soc.*, 23 (1991), 281–284.

[J3] B. E. Johnson, Permanent weak amenability of group algebras of free groups, *Bull. London Math. Soc.*, 31 (1999), 569–573.

[KR] R. V. Kadison and J. R. Ringrose, *Fundamentals of the theory of operator algebras, Volume 1, Elementary Theory*, Academic Press, New York, 1983.

[La] D. Lamb, *Beurling and Lipschitz algebras*, Thesis, University of Leeds, 1996.

[L1] A. T.-M. Lau, Operators which commute with convolutions on subspaces of $L_\infty(G)$, *Colloquium Math.*, 39 (1978), 351–359.

[L2] A. T.-M. Lau, The second conjugate algebra of the Fourier algebra of a locally compact group, *Trans. American Math. Soc.*, 267 (1981), 53–63.

[L3] A. T.-M. Lau, Continuity of Arens multiplication on the dual space of bounded uniformly continuous functions on locally compact groups and topological semigroups, *Math. Proc. Cambridge Philosophical Soc.*, 99 (1986), 273–283.

[L4] A. T.-M. Lau, Uniformly continuous functionals on Banach algebras, *Colloquium Math.*, 51 (1987), 195–205.

[LLos1] A. T.-M. Lau and V. Losert, On the second conjugate algebra of a locally compact group, *J. London Math. Soc.* (2), 37 (1988), 464–470.

[LLos2] A. T.-M. Lau and V. Losert, The C^*-algebra generated by operators with compact support on a locally compact group, *J. Functional Analysis*, 112 (1993), 1–30.

[LLo1] A. T.-M. Lau and R. J. Loy, Amenability of convolution algebras, *Mathematica Scand.*, 79 (1996), 283–296.

[LLo2] A. T.-M. Lau and R. J. Loy, Weak amenability of Banach algebras on locally compact groups, *J. Functional Analysis*, 145 (1997), 152–166.

[LPat] A. T.-M. Lau and A. L. T. Paterson, The exact cardinality of the set of topological left invariant means on an amenable locally compact group, *Proc. American Math. Soc.*, 98 (1986), 75–80.

[LPy1] A. T.-M. Lau and J. Pym, Concerning the second dual algebra of the group algebra of a locally compact group, *J. London Math. Soc.* (2), 41 (1990), 445–460.

[LPy2] A. T.-M. Lau and J. Pym, The topological centre of a compactification of a locally compact group, *Math. Zeitschrift*, 219 (1995), 567–569.

[LU] A. T.-M. Lau and A. Ülger, Topological centers of certain dual algebras, *Trans. American Math. Soc.*, 348 (1996), 1191–1212.

[LW] A. T.-M. Lau and J. C. Wong, Weakly almost periodic elements of $L_\infty(G)$ of a locally compact group, *Proc. American Math. Soc.*, 107 (1989), 1031–1036.

[LMPy] A. T.-M. Lau, A. R. Medghalchi, and J. Pym, On the spectrum of $L^\infty(G)$, *J. London Math. Soc.* (2), 48 (1993), 152–166.

[LMiPy] A. T.-M. Lau, P. Milnes, and J. Pym, Locally compact groups, invariant means and the centres of compactifications, *J. London Math. Soc.*, 56 (1997), 77–90.

[Laus] N. J. Laustsen, Maximal ideals in the algebra of operators on certain Banach spaces, *Proc. Edinburgh Math. Soc.*, 45 (2002), 523–546.

[LoW] R. J. Loy and G. A. Willis, Continuity of derivations on $\mathcal{B}(E)$ for certain Banach spaces E, *J. London Math. Soc.* (2), 40 (1989), 327–346.

[Mi1] T. Mitchell, Constant functions and left invariant means on semigroups, *Trans. American Math. Soc.*, 119 (1965), 244–261.

[Mi2] T. Mitchell, Topological semigroups and fixed points, *Illinois J. Math.*, 14 (1970), 630–641.

[N1] M. Neufang, A unified approach to the topological centre problem for certain Banach algebras arising in abstract harmonic analysis, *Archiv der Mathematik* (*Basel*), 62 (2004), 164–171.

[N2] M. Neufang, On Mazur's property and property (X), preprint, 2003.

[N3] M. Neufang, A quantized analogue of the convolution algebra $L_1(G)$, preprint, 2004.

[N4] M. Neufang, Solution to a conjecture by Ghahramani–Lau and related results on topological centres, preprint, 2004.

[N5] M. Neufang, On the topological centre problem for weighted convolution algebras and semigroup compactifications, preprint, 2004.

[Ol1] A. Yu. Ol'shanskii, An infinite group with subgroups of prime orders, *Izvestiya Akademic Nauk SSSR, Ser. Mat.*, 44 (1980), 309–321.

[Ol2] A. Yu. Ol'shanskii, On the problem of the existence of an invariant mean on a group, *Uspekhi Mat. Nauk*, 35 (4), (1980), 199–200 = *Russian Math. Surveys*, 35 (4), (1980), 180–181.

[Pal] T. W. Palmer, The bidual of the compact operators, *Trans. American Math. Soc.*, 288 (1985), 827–839.

[Pa2] T. W. Palmer, *Banach algebras and the general theory of ⋆-algebras, Volume 1, Algebras and Banach algebras*, Cambridge University Press, 1994.

[Pa3] T. W. Palmer, *Banach algebras and the general theory of ⋆-algebras, Volume 2*, Cambridge University Press, 2001.

[Par] D. J. Parsons, The centre of the second dual of a commutative semigroup algebra, *Math. Proc. Cambridge Philosophical Soc.*, 95 (1984), 71–92.

[Pat1] A. L. T. Paterson, Amenable groups for which every topological left invariant mean is invariant, *Pacific J. Math.*, 84 (1979), 391–397.

[Pat2] A. L. T. Paterson, *Amenability*, American Math. Soc., Providence, Rhode Island, 1988.

[Pi] J. Pier, *Amenable locally compact groups*, Wiley, New York, 1984.

[Pir] A. Yu. Pirkovskii, Biprojectivity and biflatness for convolution algebras of nuclear operators, *Canadian Math. Bulletin*, 47 (2004), 445–455.

[Py1] J. S. Pym, The convolution of functionals on spaces of bounded functions, *Proc. London Math. Soc.* (3), 15 (1965), 84–104.

[Py2] J. S. Pym, Remarks on the second duals of Banach algebras, *J. Nigerian Math. Soc.*, 2 (1983), 31–33.

[PyU] J. S. Pym and A. Ülger, On the Arens regularity of inductive limit algebras and related matters, *Quarterly J. Math. Oxford* (2), 40 (1989), 101–109.

[RS] H. Reiter and J. D. Stegeman, *Classical harmonic analysis and locally compact groups*, London Math. Society Monographs, Volume 22, Clarendon Press, Oxford, 2000.

[Ri] C. E. Rickart, *General theory of Banach algebras*, D. van Nostrand, Princeton, New Jersey, 1960.

[Ru1] V. Runde, Amenability for dual Banach algebras, *Studia Mathematica*, 148 (2001), 47–66.

[Ru2] V. Runde, *Lectures on amenability*, Lecture Notes in Mathematics, Volume 1774, Springer–Verlag, Berlin, 2002.

[Rup] W. Ruppert, *Compact semitopological semigroups: an intrinsic theory*, Lecture Notes in Mathematics, Volume 1079, Springer–Verlag, Berlin, 1984.

[Sh] S. Sherman, The second adjoint of a C^*-algebra, *Proc. International Congress Mathematicians*, Cambridge, Mass., Volume 1, (1950), 470.

[Si] A. I. Singh, $L_0^\infty(G)^*$ as the second dual of the group algebra $L^1(G)$ with a locally convex topology, *Michigan Math. J.*, 46(1999), 143–150.

[Td] Z. Takeda, Conjugate spaces of operator algebras, *Proc. Japan Academy*, 30 (1954), 90–95.

[Ta] M. Takesaki, *Theory of operator algebras, Volume 1*, Springer-Verlag, Berlin, 1979.

[U1] A. Ülger, Arens regularity of the algebra $A \widehat{\otimes} B$, *Trans. American Math. Soc.*, 305 (1988), 623–639.

[U2] A. Ülger, Arens regularity of the algebra $C(K, A)$, *J. London Math. Soc.* (2), 42 (1990), 354–364.

[U3] A. Ülger, Some stability properties of Arens regular bilinear operators, *Proc. Edinburgh Math. Soc.* (2), 34 (1991), 443–454.

[U4] A. Ülger, Arens regularity of weakly sequentially complete Banach algebras, *Proc. American Math. Soc.*, 127 (1999), 3221–3227.

[U5] A. Ülger, Central elements of A^{**} for certain Banach algebras A without bounded approximate identities, *Glasgow Math. Journal*, 41 (1999), 369–377.

[We] J. G. Wendel, Left centralizers and isomorphisms of group algebras, *Pacific J. Math.*, 2 (1952), 251–261.

[Wh] M. C. White, Characters on weighted amenable groups, *Bull. London Math. Soc.*, 23 (1991), 375–380.

[Wo] J. C. S. Wong, Topologically stationary locally compact groups and amenability, *Trans. American Math. Soc.*, 144 (1969), 351–363.

[Y1] N. J. Young, Separate continuity and multilinear operations, *Proc. London Math. Soc.* (3), 26 (1973), 289–319.

[Y2] N. J. Young, The irregularity of multiplication in group algebras, *Quarterly J. Math. Oxford* (2), 24 (1973), 59–62.

[Y3] N. J. Young, Periodicity of functionals and representations of normed algebras on reflexive spaces, *Proc. Edinburgh Math. Soc.* (2), 20 (1976), 99–120.

[Z] A. Zappa, The centre of the convolution algebra $C_u(G)^*$, *Rend. Sem. Mat. Univ. Padova*, 52 (1974), 71–83.

Index

algebra,
  Banach $*$-, 13
  Banach operator, 54
  C^*-, 13, 39
  Fourier, 43
  Hertz, 43
  multiplier, 9
  nuclear, 54
  radical, 8
  semisimple, 8, 175
  Volterra, 42
  von Neumann, 13
    enveloping, 37
  $*$-semisimple, 9
almost periodic function, 69
almost periodic functional, 32
amenable Banach algebra, 15
amenable locally compact group, 89
annihilator, 12
approximation property, 59
Arens products, first and second, 16
Arens regular, 1, 16
Arens, Richard, 1
augmentation
  character, 75
  ideal, 75

Banach A-bimodule, 14
  essential, 14
Banach algebra,
  2-weakly amenable, 15, 167
  amenable, 15
  dual, 15
  operator algebra, 54
  weakly amenable, 15
Banach left, right A-module, 14
Banach space,
  reflexive, 11
  super-reflexive, 56
  uniformly convex, 56
  weakly sequentially complete, 66
Beurling algebra,
  continuous, 73
  discrete, 71
bimodule, 10
  neo-unital, 10
bounded approximate identity, 11
  sequential, 11
bounded compact approximation property, 54

C^*-algebra, 13, 37
canonical projection, 11, 15
canonical representation, 131
centre, 7
character, 10
  augmentation, 75
clusters, 26,27
  0-, 26,27
  0-, locally uniformly, 30, 157
  0-, strongly, 30, 31
compactification,
  AP-, 82
  Bohr, 82
  of a group, 159
  LUC-, 82
  of semigroups, 82
  Stone-Čech, 13, 81
  WAP-, 82
components, 131
convolution product, 73

decomposable, 8
  strongly, 10

derivation, 10, 167
  inner, 10
diagonally bounded, 92
dispersed subset, 147
dual, 53

Feinstein example, 128, 175, 176
Fourier algebra, 43
Fourier transform, 72
function,
  almost periodic, 69
  left uniformly continuous, 68
  right uniformly continuous, 68
  symmetric, 70
  weakly almost periodic, 69

group,
  Banach cohomology, 15
  Cantor, 36
  [FC], 89
  free, 76, 131
  locally compact, 38
    amenable, 89
  maximally almost periodic, 76
group algebra, 38

Haar measure, 65
hermitian, 97
Hertz algebra, 43

introverted, 45, 47
invariant, 86
  right-, 86
  almost, 90
  almost left-, 90
  left-, 86
  left-S-, 86
  left-s-, 86
  topologically left-, 86
  translation-, 141
involution, 9
  linear, 9

James space, 42, 63

left-annihilator, 8
linear functional,
  almost periodic, 32
  multiplicative, 109
  positive, 9
  weakly almost periodic, 32

Mazur's property, 155
mean, 86
measure algebra, 39
mixed identity, 19
modular function, 65
module,
  dual, 15
  faithful, 45
  introverted, 47
    left-, 45, 141
    right-, 47
  second dual, 14
  symmetric, 9
multiplier, 9
  algebra, 9, 84
  left, right, 8

nilpotent, 7
  of index n, 8
nuclear algebra, 54

operator,
  compact, 12
  approximable, 53
  integral, 59
  nuclear, 54
  weakly compact, 12

period, 162
point mass, normalized, 96
property (X), 155

quasi-nilpotent, 10

radical, 8, 83
Radon–Nikodým property, 59
regular, Arens 3, 16
repeated limits, 25

semidirect product, 8, 20
semisimple, 8
splitting homomorphism, 10, 20
state, 9
  pure, 13
state space, 9, 96
Stone-Čech compactification, 13, 81
strongly Arens irregular, 1, 22
  left, right 22

support, 5

tensor product,
  injective, 53
  projective, 12, 53
theorem
  Bade, 147
  Cohen's factorization, 14
  Grønbæk, 77
  Grothendieck, 25, 27, 59
  Steinhaus, 66
  Wendel, 76
thick, left, 108
topological centre, 3, 21, 47, 82
translation, 68

ultrafilter, 56
ultrapower, 56

Volterra algebra, 42
von Neumann algebra, 13

weakly almost periodic function, 68
weakly almost periodic functional, 32
weakly unconditionally Cauchy, 154
weight, 70
  almost multiplicative, 109
  diagonally bounded, 92, 157
weight function, 72
weight functions, equivalent, 74

Index of Symbols

A^{op}, 7
$A^{\#}$, 7
$A \otimes B$, 9
$A(G)$, 43
$A(X)$, 72
$AP(A)$, 32
$AP(G)$, 69
$AP(G, 1/\omega)$, 70
$A \cdot E$, AE, 10
$\mathcal{A}(E)$, 53
$A_p(G)$, $A_p(E)$, 43
$A_\omega(S)$, 96
A_ω, A'_ω, 95, 111
$A^+_\omega, (A'_\omega)^+$, 97
$\mathcal{A}_\omega$, $\mathcal{A}'_\omega$, 78
$\mathcal{A}P_\omega$, 78
$\mathrm{ac}\, S$, 7

$B_\omega(S)$, 96
B_ω, 111
B^+_ω, 97
$\mathcal{B}_\omega$, 78
$\mathcal{B}(E, F), \mathcal{B}(E)$, 12
$\mathcal{B}(E)^a$, 53
$\beta\Omega$, 13

$C_0(\Omega)$, 13
$CB(G)$, 68
$CB(G, 1/\omega)$, 70
$C_0(G, 1/\omega)$, 66
$c_0(S, 1/\omega)$, 66
$c_0(A_n)$, 35
$C_\star$, 42

$\mathbb{D}(z; r)$, $\mathbb{D}$, 5
Δ_G, 65
Δ_ω, 96
δ_x, 10
δ_s, 65
$(\delta^1\eta)(s, t)$, 71

$E^\times$, 7
E', 11
E_S, 96
$E_{[m]}$, 11
$(E \widehat{\otimes} F, \|\cdot\|_\pi)$, 12
$E\check{\otimes}E'$, $E\widehat{\otimes}E'$, 53
E_ω, 95
$E^{\circ+}_\omega, E^{\circ-}_\omega$, 104
$\mathcal{E}_\omega$, 78
$\mathcal{E}^\circ_\omega$, 83
$\mathrm{ex}\, S$, 7

$\mathbb{F}_2$, 131
$F_{\mathcal{U}}$, 56

$\check{f}$, 74
f^*, 75
$\widehat{f}$, 46
F°, 12
$\Phi^\triangleleft$, 97
Φ_A, 10
$\Phi_\ell(\lambda), \Phi_r(\lambda)$, 141
φ_G, 75

${}^\circ G$, 12
$H^1(\mathbb{T})$, 43
$\mathcal{H}^1(A, E)$, 15

$\mathbb{I}$, 5
$I_\lambda(S)$, 107
I_m, 107
$\mathcal{I}(E)$, 59
$\operatorname{Inv} A$, 8

K_λ, 141
$\mathcal{K}(E, F), \mathcal{K}(E)$, 12
κ_E, 11
$\kappa(\Omega)$, 91

L_a, 7
$LO(\lambda)$, 68
$LUC(G)$, 68
$LUC(G, 1/\omega)$, 70
$(L^1(G), \star)$, 40
$(L^1(\mathbb{I}), \star)$, 44
$L^1(G,\omega), L^\infty(G, 1/\omega)$, 66, 73
$L^1_{\mathbb{R}}(G,\omega), L^\infty_{\mathbb{R}}(G, 1/\omega)$, 66
$L^2(G)$, 75
$L^p(\mu)$, 56
$(L^1(\mathbb{R}^+,\omega), \star)$, 42
$L^\infty_{00}(G, 1/\omega)$, 66
$\ell^1(G,\omega)$, 71
$\ell^1(S,\omega), \ell^\infty(S, 1/\omega)$, 65
$\ell^2(E)$, 55
$\ell^\infty(A_k)$, 111
ℓ_t, r_t, 68
$\mathcal{L}(E, F)$, $\mathcal{L}(E)$, 7
$\mathcal{L}_{t,\omega}(G), \mathcal{L}_t(G)$, 89

$\lim_m \lim_n f(s_m, t_n)$, 25
$\operatorname{Lim}_\alpha K_\alpha$, 5
$\operatorname{Lim}_\alpha s_\alpha$, 5
$\operatorname{Lim}_{x\to\infty} \operatorname{Lim}_{y\to\infty} f(x, y)$, 25
λ_s, 65
$\lambda^\triangleleft$, 18

$\mathbb{M}_n$, 5
$M(\Omega)$, 13
$M(G)$, 39
$M(G,\omega)$, $M_a(G,\omega), M_s(G,\omega)$, 67
m_A, 7
$\mathcal{M}(A)$, 9
$\mathcal{M}_\omega$, 78
$\mu^*(E)$, 75

$\mathbb{N}$, $\mathbb{N}_k$, 5
$\mathcal{N}^1(A, E)$, 15
$\nu(a)$, 10

$P(G), P_\omega(G)$, 66
Π_ω, 85

$\mathfrak{Q}(A)$, 10

R_a, 7
$RUC(G)$, 68
$RUC(G, 1/\omega)$, 70
$\mathcal{R}^\square_\omega, \mathcal{R}^\diamond_\omega$, 83
$R^\square_\omega, R^\diamond_\omega$, 95
$\operatorname{rad} A$, 8

$S + T$, $S \cdot T$, $S + \{t\}$, 5
S^{-1}, $S^\bullet$, 5
$S^{[n]}, S^n$, $\langle S\rangle$, 7
S_A, 9
S_ω, 96
$\mathcal{S}_\omega$, 78
$\sigma(E'', E')$, 11
$\sigma(a)$, 10
$\operatorname{supp} f$, 5

$\mathbb{T}$, 5

T', 53

$WAP(A)$, 32
$WAP(G)$, 69
$WAP(G, 1/\omega)$, 70
$\Omega(s,t)$, 71
$\Omega_k(s_1, \ldots, s_k)$, 102
ω_α, 72,73

$X(\omega)$, 66
$\mathcal{X}_\omega$, 78
$\mathcal{X}'_\omega$, $\mathcal{X}^\circ_\omega$, 83

$\mathcal{W}_\omega$, 78

$\mathcal{W}(E, F), \mathcal{W}(E)$, 12

$\mathbb{Z}^+$, $\mathbb{Z}^+_k$, 5
$\mathcal{Z}^1(A, E)$, 15
$\mathfrak{Z}_t^{(1)}(A'')$, $\mathfrak{Z}_t^{(2)}(A'')$, 21
$\mathfrak{Z}_t(X')$, 47, 48
$\mathfrak{Z}_t(\mathcal{X}'_\omega)$, 85
$\mathfrak{Z}(B_\omega)$, 111

$\ltimes$, 8
$\star$, 71, 73, 75
$\cdot_\omega$, 67, 96
$\Box$, $\Diamond$, 16
$\langle \cdot, \cdot \rangle$, 7, 77
$\| \cdot \|_{\mathcal{N}}$, 54

Editorial Information

To be published in the *Memoirs*, a paper must be correct, new, nontrivial, and significant. Further, it must be well written and of interest to a substantial number of mathematicians. Piecemeal results, such as an inconclusive step toward an unproved major theorem or a minor variation on a known result, are in general not acceptable for publication. Papers appearing in *Memoirs* are generally at least 80 and not more than 200 published pages in length. Papers less than 80 or more than 200 published pages require the approval of the Managing Editor of the Transactions/Memoirs Editorial Board.

As of May 31, 2005, the backlog for this journal was approximately 11 volumes. This estimate is the result of dividing the number of manuscripts for this journal in the Providence office that have not yet gone to the printer on the above date by the average number of monographs per volume over the previous twelve months, reduced by the number of volumes published in four months (the time necessary for preparing a volume for the printer). (There are 6 volumes per year, each containing at least 4 numbers.)

A Consent to Publish and Copyright Agreement is required before a paper will be published in the *Memoirs*. After a paper is accepted for publication, the Providence office will send a Consent to Publish and Copyright Agreement to all authors of the paper. By submitting a paper to the *Memoirs*, authors certify that the results have not been submitted to nor are they under consideration for publication by another journal, conference proceedings, or similar publication.

Information for Authors

Memoirs are printed from camera copy fully prepared by the author. This means that the finished book will look exactly like the copy submitted.

The paper must contain a *descriptive title* and an *abstract* that summarizes the article in language suitable for workers in the general field (algebra, analysis, etc.). The *descriptive title* should be short, but informative; useless or vague phrases such as "some remarks about" or "concerning" should be avoided. The *abstract* should be at least one complete sentence, and at most 300 words. Included with the footnotes to the paper should be the 2000 *Mathematics Subject Classification* representing the primary and secondary subjects of the article. The classifications are accessible from `www.ams.org/msc/`. The list of classifications is also available in print starting with the 1999 annual index of *Mathematical Reviews*. The Mathematics Subject Classification footnote may be followed by a list of *key words and phrases* describing the subject matter of the article and taken from it. Journal abbreviations used in bibliographies are listed in the latest *Mathematical Reviews* annual index. The series abbreviations are also accessible from `www.ams.org/publications/`. To help in preparing and verifying references, the AMS offers MR Lookup, a Reference Tool for Linking, at `www.ams.org/mrlookup/`. When the manuscript is submitted, authors should supply the editor with electronic addresses if available. These will be printed after the postal address at the end of the article.

Electronically prepared manuscripts. The AMS encourages electronically prepared manuscripts, with a strong preference for $\mathcal{A}_{\mathcal{M}}\mathcal{S}$-LaTeX. To this end, the Society has prepared $\mathcal{A}_{\mathcal{M}}\mathcal{S}$-LaTeX author packages for each AMS publication. Author packages include instructions for preparing electronic manuscripts, the *AMS Author Handbook*, samples, and a style file that generates the particular design specifications of that publication series. Though $\mathcal{A}_{\mathcal{M}}\mathcal{S}$-LaTeX is the highly preferred format of TeX, author packages are also available in $\mathcal{A}_{\mathcal{M}}\mathcal{S}$-TeX.

Authors may retrieve an author package from e-MATH starting from `www.ams.org/tex/` or via FTP to `ftp.ams.org` (login as `anonymous`, enter username as password, and type `cd pub/author-info`). The *AMS Author Handbook* and the *Instruction Manual* are available in PDF format following the author packages link from `www.ams.org/tex/`. The author package can be obtained free of charge by sending email to pub@ams.org (Internet) or from the Publication Division, American Mathematical Society, 201 Charles St., Providence, RI 02904, USA. When requesting an author package, please specify $\mathcal{A}_\mathcal{M}\mathcal{S}$-LaTeX or $\mathcal{A}_\mathcal{M}\mathcal{S}$-TeX, Macintosh or IBM (3.5) format, and the publication in which your paper will appear. Please be sure to include your complete mailing address.

Sending electronic files. After acceptance, the source file(s) should be sent to the Providence office (this includes any TeX source file, any graphics files, and the DVI or PostScript file).

Before sending the source file, be sure you have proofread your paper carefully. The files you send must be the EXACT files used to generate the proof copy that was accepted for publication. For all publications, authors are required to send a printed copy of their paper, which exactly matches the copy approved for publication, along with any graphics that will appear in the paper.

TeX files may be submitted by email, FTP, or on diskette. The DVI file(s) and PostScript files should be submitted only by FTP or on diskette unless they are encoded properly to submit through email. (DVI files are binary and PostScript files tend to be very large.)

Electronically prepared manuscripts can be sent via email to pub-submit@ams.org (Internet). The subject line of the message should include the publication code to identify it as a Memoir. TeX source files, DVI files, and PostScript files can be transferred over the Internet by FTP to the Internet node e-math.ams.org (130.44.1.100).

Electronic graphics. Comprehensive instructions on preparing graphics are available at www.ams.org/jourhtml/graphics.html. A few of the major requirements are given here.

Submit files for graphics as EPS (Encapsulated PostScript) files. This includes graphics originated via a graphics application as well as scanned photographs or other computer-generated images. If this is not possible, TIFF files are acceptable as long as they can be opened in Adobe Photoshop or Illustrator. No matter what method was used to produce the graphic, it is necessary to provide a paper copy to the AMS.

Authors using graphics packages for the creation of electronic art should also avoid the use of any lines thinner than 0.5 points in width. Many graphics packages allow the user to specify a "hairline" for a very thin line. Hairlines often look acceptable when proofed on a typical laser printer. However, when produced on a high-resolution laser imagesetter, hairlines become nearly invisible and will be lost entirely in the final printing process.

Screens should be set to values between 15% and 85%. Screens which fall outside of this range are too light or too dark to print correctly. Variations of screens within a graphic should be no less than 10%.

Inquiries. Any inquiries concerning a paper that has been accepted for publication should be sent directly to the Electronic Prepress Department, American Mathematical Society, 201 Charles St., Providence, RI 02904, USA.

Editors

Titles in This Series

836 **H. G. Dales and A. T.-M. Lau,** The second duals of Beurling algebras, 2005

835 **Kiyoshi Igusa,** Higher complex torsion and the framing principle, 2005

834 **Ken'ichi Ohshika,** Kleinian groups which are limits of geometrically finite groups, 2005

833 **Greg Hjorth and Alexander S. Kechris,** Rigidity theorems for actions of product groups and countable Borel equivalence relations, 2005

832 **Lee Klingler and Lawrence S. Levy,** Representation type of commutative Noetherian rings III: Global wildness and tameness, 2005

831 **K. R. Goodearl and F. Wehrung,** The complete dimension theory of partially ordered systems with equivalence and orthogonality, 2005

830 **Jason Fulman, Peter M. Neumann, and Cheryl E. Praeger,** A generating function approach to the enumeration of matrices in classical groups over finite fields, 2005

829 **S. G. Bobkov and B. Zegarlinski,** Entropy bounds and isoperimetry, 2005

828 **Joel Berman and Paweł M. Idziak,** Generative complexity in algebra, 2005

827 **Trevor A. Welsh,** Fermionic expressions for minimal model Virasoro characters, 2005

826 **Guy Métivier and Kevin Zumbrun,** Large viscous boundary layers for noncharacteristic nonlinear hyperbolic problems, 2005

825 **Yaozhong Hu,** Integral transformations and anticipative calculus for fractional Brownian motions, 2005

824 **Luen-Chau Li and Serge Parmentier,** On dynamical Poisson groupoids I, 2005

823 **Claus Mokler,** An analogue of a reductive algebraic monoid whose unit group is a Kac-Moody group, 2005

822 **Stefano Pigola, Marco Rigoli, and Alberto G. Setti,** Maximum principles on Riemannian manifolds and applications, 2005

821 **Nicole Bopp and Hubert Rubenthaler,** Local zeta functions attached to the minimal spherical series for a class of symmetric spaces, 2005

820 **Vadim A. Kaimanovich and Mikhail Lyubich,** Conformal and harmonic measures on laminations associated with rational maps, 2005

819 **F. Andreatta and E. Z. Goren,** Hilbert modular forms: Mod p and p-adic aspects, 2005

818 **Tom De Medts,** An algebraic structure for Moufang quadrangles, 2005

817 **Javier Fernández de Bobadilla,** Moduli spaces of polynomials in two variables, 2005

816 **Francis Clarke,** Necessary conditions in dynamic optimization, 2005

815 **Martin Bendersky and Donald M. Davis,** V_1-periodic homotopy groups of $SO(n)$, 2004

814 **Johannes Huebschmann,** Kähler spaces, nilpotent orbits, and singular reduction, 2004

813 **Jeff Groah and Blake Temple,** Shock-wave solutions of the Einstein equations with perfect fluid sources: Existence and consistency by a locally inertial Glimm scheme, 2004

812 **Richard D. Canary and Darryl McCullough,** Homotopy equivalences of 3-manifolds and deformation theory of Kleinian groups, 2004

811 **Ottmar Loos and Erhard Neher,** Locally finite root systems, 2004

810 **W. N. Everitt and L. Markus,** Infinite dimensional complex symplectic spaces, 2004

809 **J. T. Cox, D. A. Dawson, and A. Greven,** Mutually catalytic super branching random walks: Large finite systems and renormalization analysis, 2004

808 **Hagen Meltzer,** Exceptional vector bundles, tilting sheaves and tilting complexes for weighted projective lines, 2004

807 **Carlos A. Cabrelli, Christopher Heil, and Ursula M. Molter,** Self-similarity and multiwavelets in higher dimensions, 2004

806 **Spiros A. Argyros and Andreas Tolias,** Methods in the theory of hereditarily indecomposable Banach spaces, 2004

805 **Philip L. Bowers and Kenneth Stephenson,** Uniformizing dessins and Belyĭ maps via circle packing, 2004

804 **A. Yu Ol'shanskii and M. V. Sapir,** The conjugacy problem and Higman embeddings, 2004

803 **Michael Field and Matthew Nicol,** Ergodic theory of equivariant diffeomorphisms: Markov partitions and stable ergodicity, 2004

802 **Martin W. Liebeck and Gary M. Seitz,** The maximal subgroups of positive dimension in exceptional algebraic groups, 2004

801 **Fabio Ancona and Andrea Marson,** Well-posedness for general 2×2 systems of conservation law, 2004

800 **V. Poénaru and C. Tanas,** Equivariant, almost-arborescent representation of open simply-connected 3-manifolds; A finiteness result, 2004

799 **Barry Mazur and Karl Rubin,** Kolyvagin systems, 2004

798 **Benoît Mselati,** Classification and probabilistic representation of the positive solutions of a semilinear elliptic equation, 2004

797 **Ola Bratteli, Palle E. T. Jorgensen, and Vasyl' Ostrovs'kyĭ,** Representation theory and numerical AF-invariants, 2004

796 **Marc A. Rieffel,** Gromov-Hausdorff distance for quantum metric spaces/Matrix algebras converge to the sphere for quantum Gromov-Hausdorff distance, 2004

795 **Adam Nyman,** Points on quantum projectivizations, 2004

794 **Kevin K. Ferland and L. Gaunce Lewis, Jr.,** The $RO(G)$-graded equivariant ordinary homology of G-cell complexes with even-dimensional cells for $G = \mathbb{Z}/p$, 2004

793 **Jindřich Zapletal,** Descriptive set theory and definable forcing, 2004

792 **Inmaculada Baldomá and Ernest Fontich,** Exponentially small splitting of invariant manifolds of parabolic points, 2004

791 **Eva A. Gallardo-Gutiérrez and Alfonso Montes-Rodríguez,** The role of the spectrum in the cyclic behavior of composition operators, 2004

790 **Thierry Lévy,** Yang-Mills measure on compact surfaces, 2003

789 **Helge Glöckner,** Positive definite functions on infinite-dimensional convex cones, 2003

788 **Robert Denk, Matthias Hieber, and Jan Prüss,** $\mathcal{R}$-boundedness, Fourier multipliers and problems of elliptic and parabolic type, 2003

787 **Michael Cwikel, Per G. Nilsson, and Gideon Schechtman,** Interpolation of weighted Banach lattices/A characterization of relatively decomposable Banach lattices, 2003

786 **Arnd Scheel,** Radially symmetric patterns of reaction-diffusion systems, 2003

785 **R. R. Bruner and J. P. C. Greenlees,** The connective K-theory of finite groups, 2003

784 **Desmond Sheiham,** Invariants of boundary link cobordism, 2003

783 **Ethan Akin, Mike Hurley, and Judy A. Kennedy,** Dynamics of topologically generic homeomorphisms, 2003

782 **Masaaki Furusawa and Joseph A. Shalika,** On central critical values of the degree four L-functions for GSp(4): The Fundamental Lemma, 2003

781 **Marcin Bownik,** Anisotropic Hardy spaces and wavelets, 2003

780 **S. Marmi and D. Sauzin,** Quasianalytic monogenic solutions of a cohomological equation, 2003

779 **Hansjörg Geiges,** h-principles and flexibility in geometry, 2003

778 **David B. Massey,** Numerical control over complex analytic singularities, 2003

For a complete list of titles in this series, visit the
AMS Bookstore at **www.ams.org/bookstore/**.